Charles Lewis Strobel

Pocket Companion of Useful Information and Tables

Charles Lewis Strobel

Pocket Companion of Useful Information and Tables

ISBN/EAN: 9783337398958

Printed in Europe, USA, Canada, Australia, Japan

Cover: Foto ©Suzi / pixelio.de

More available books at **www.hansebooks.com**

SHAPES.

MANUFACTURED BY

CARNEGIE BROS. & CO.

LIMITED

PROPRIETORS

UNION IRON MILLS.

Pittsburgh, Pa.

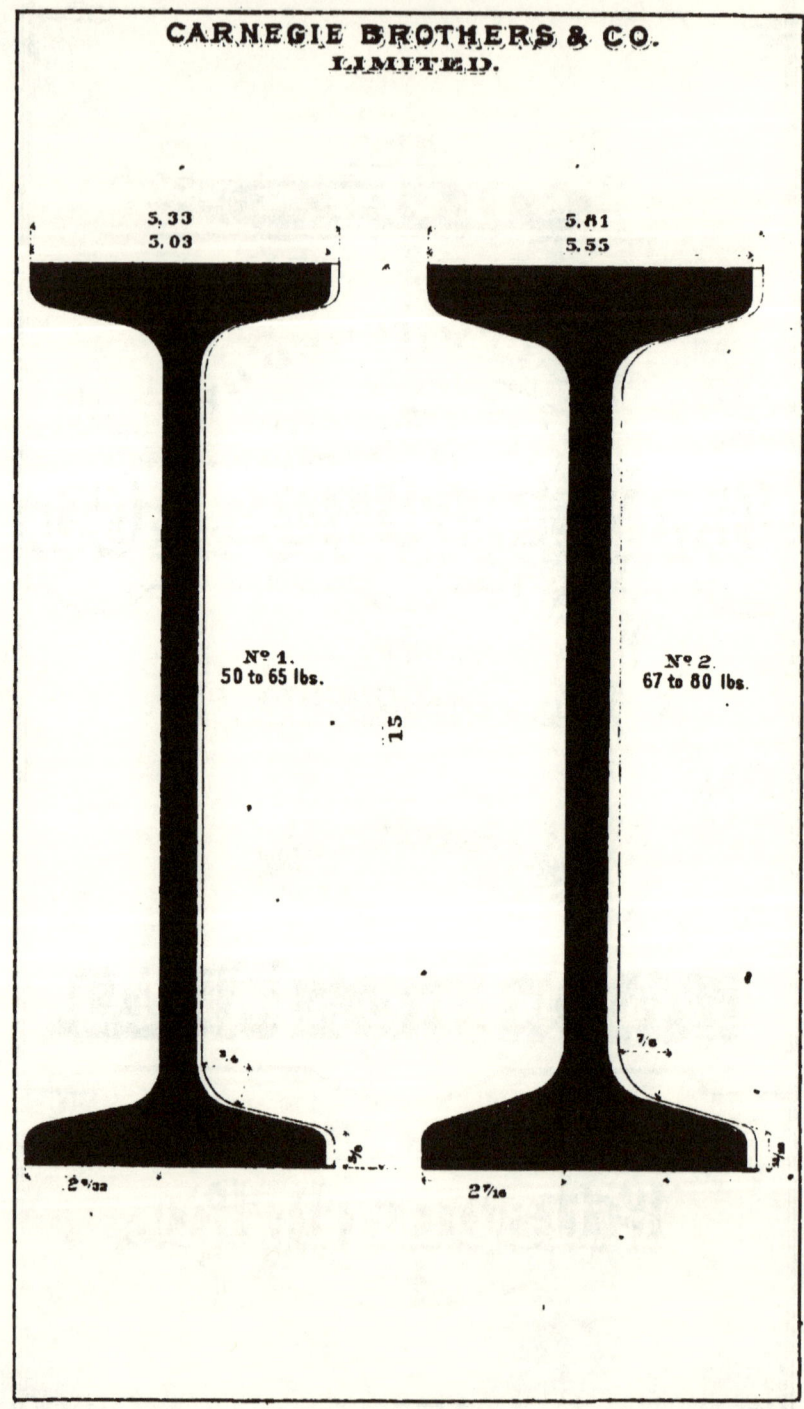

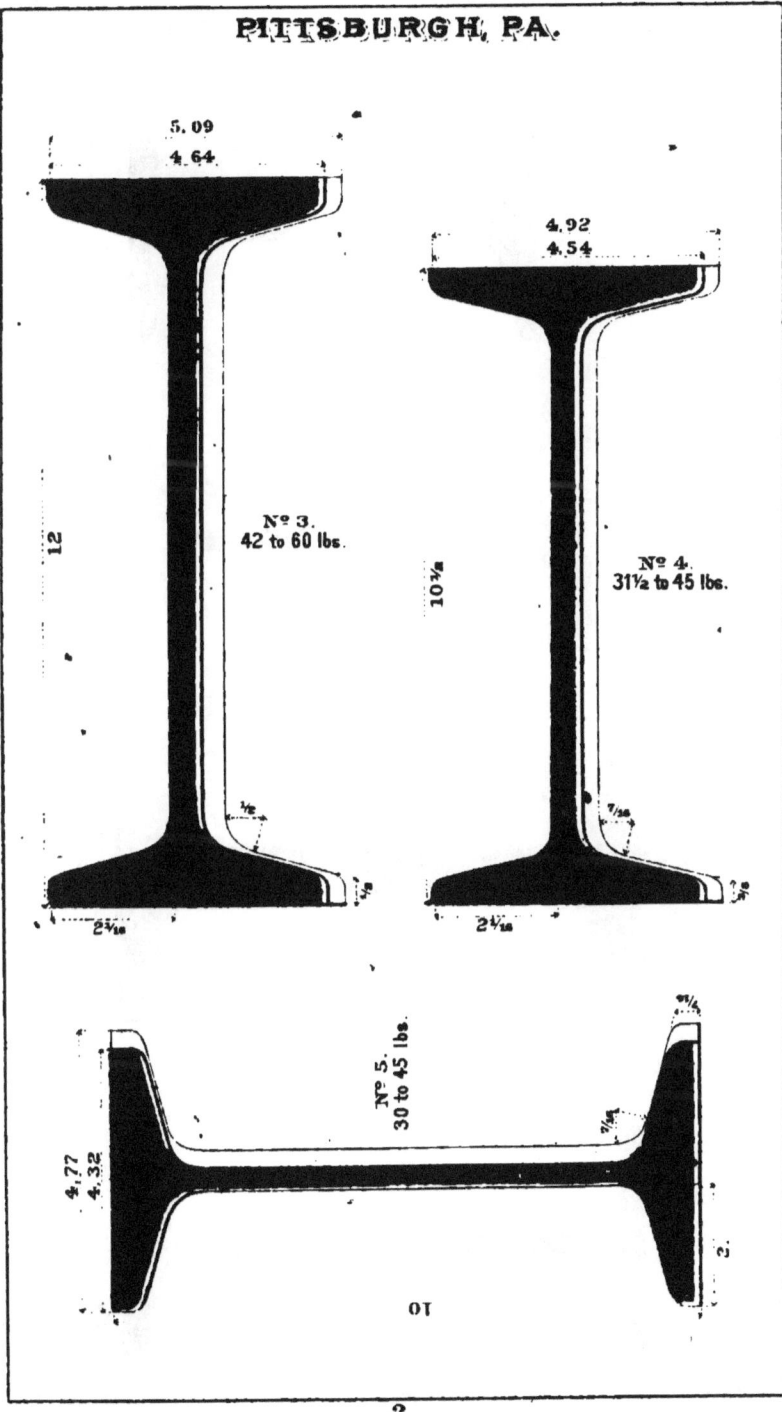

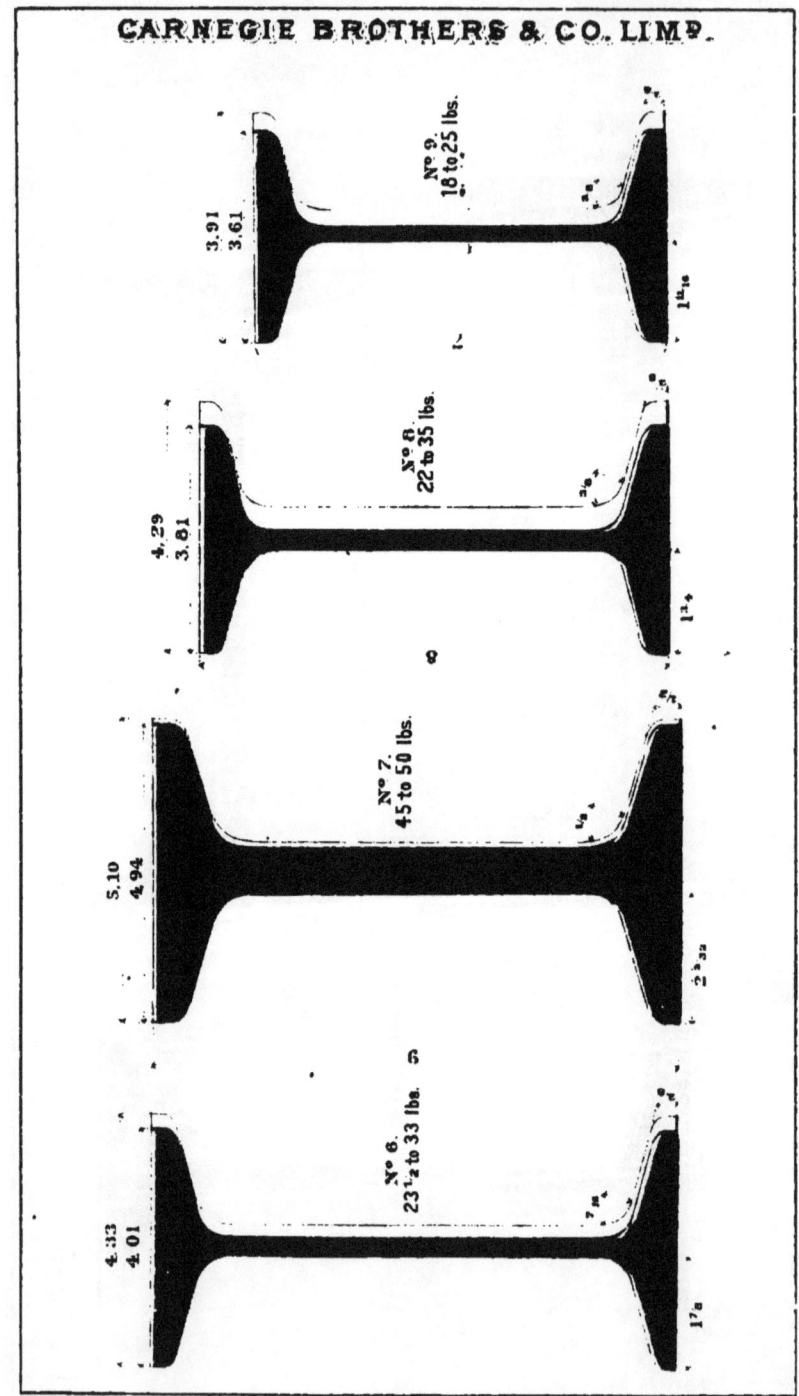

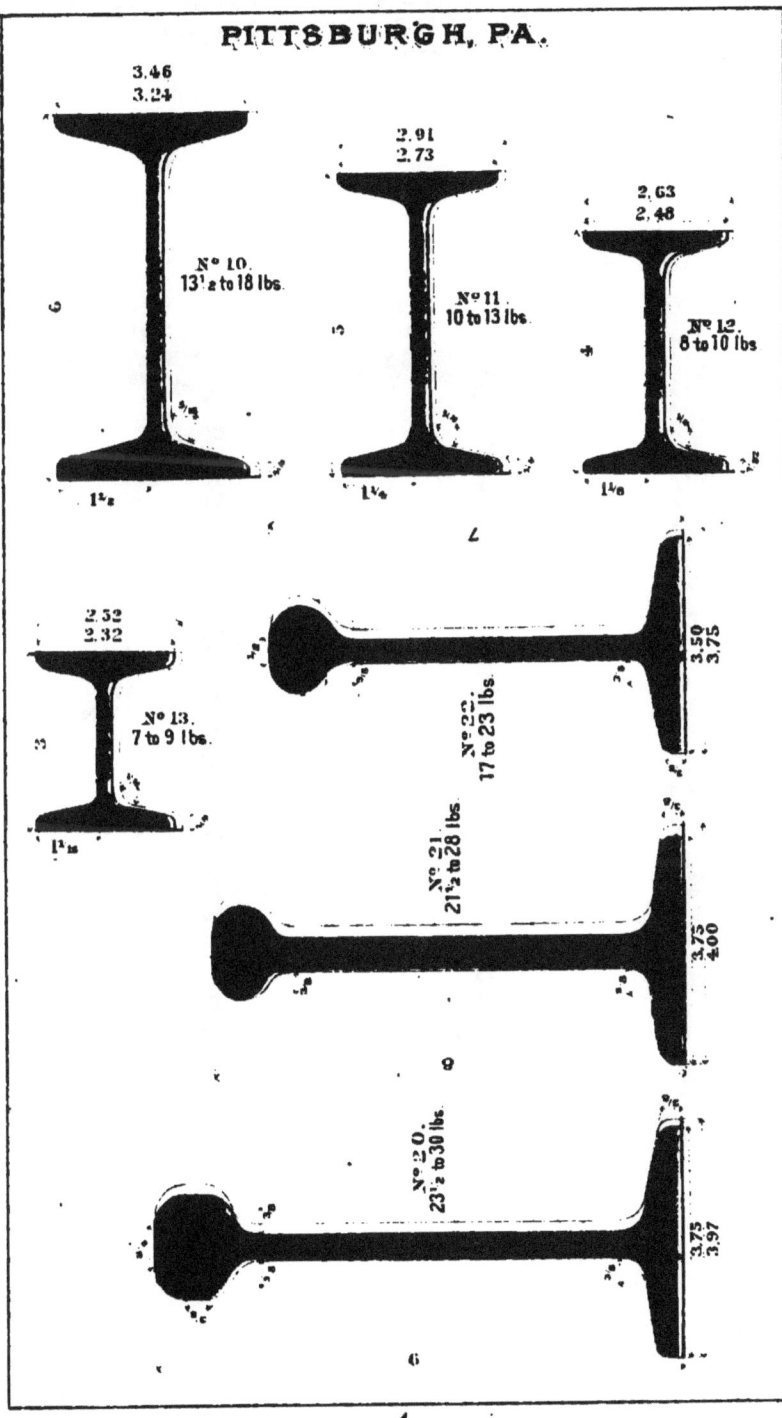

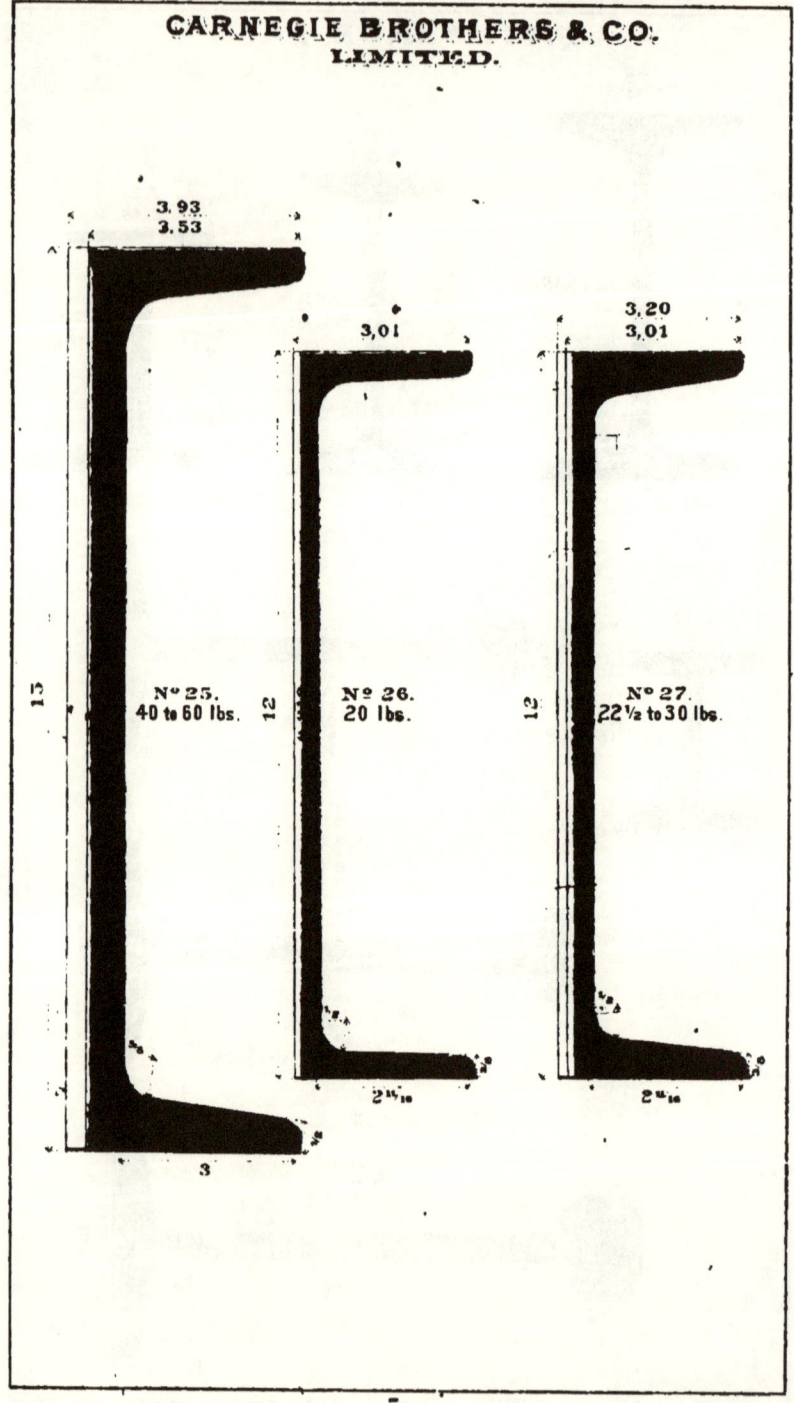

PITTSBURGH, PA.

N° 28. 30 to 50 lbs.

N° 29. 16 lbs.

N° 30. 17½ to 30 lbs.

N° 32. 14½ lbs.

N° 31. 20 to 35 lbs.

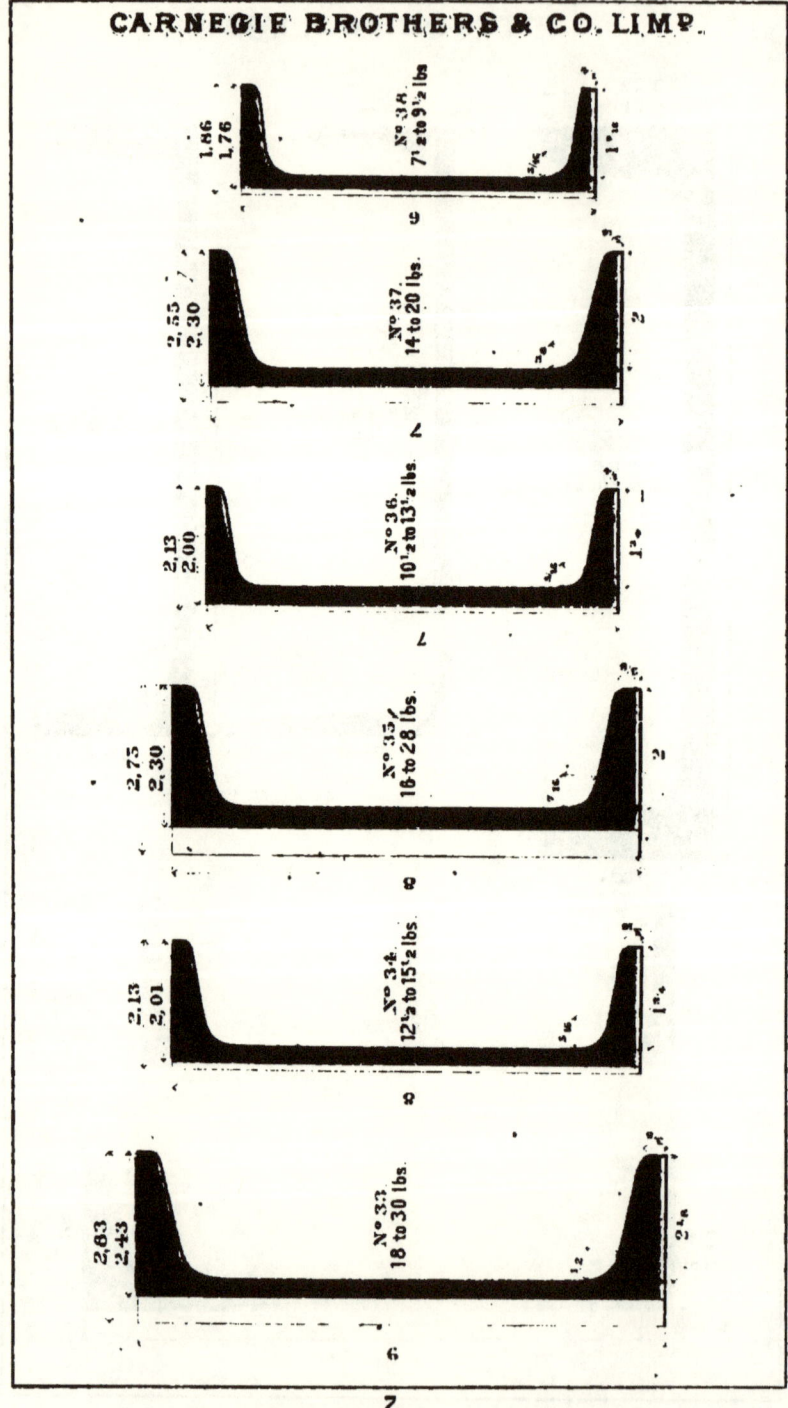

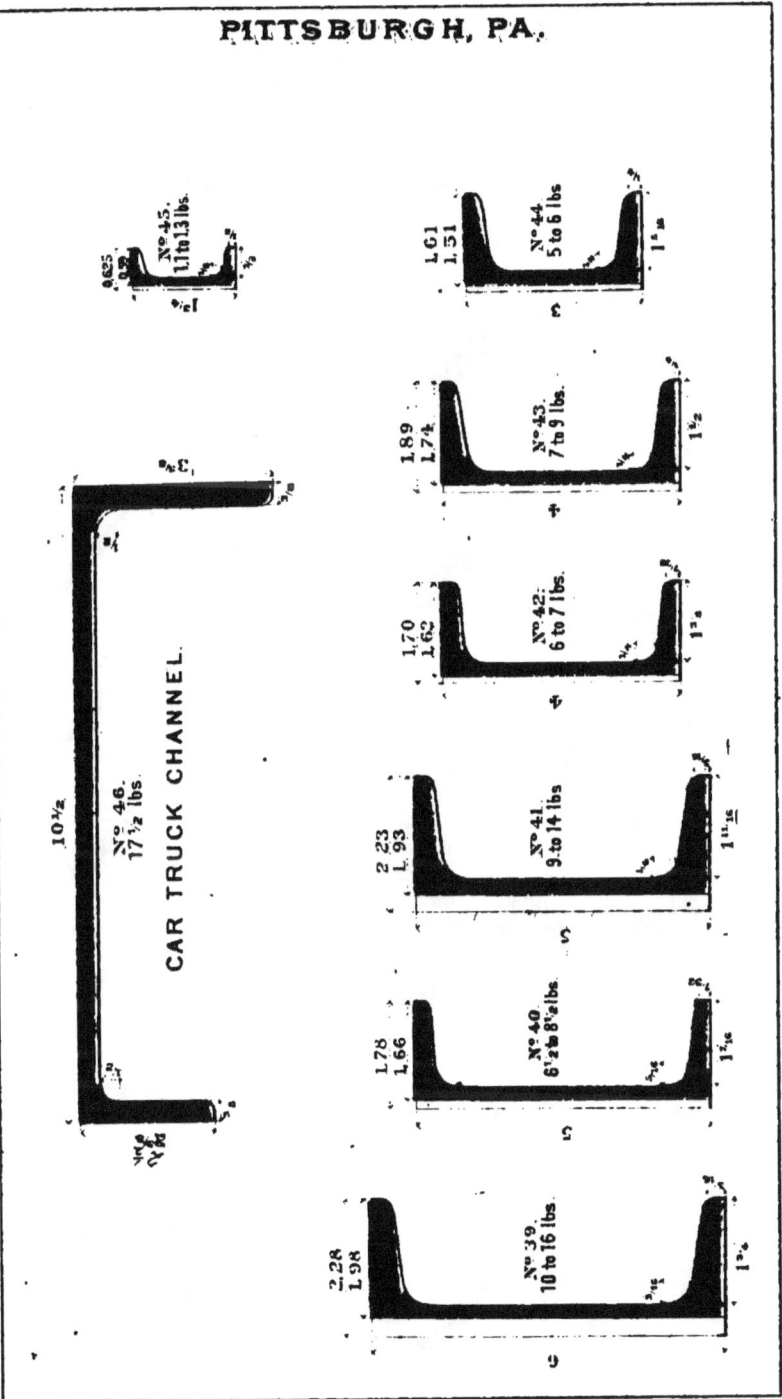

CARNEGIE BROTHERS & CO., LIMITED.

ANGLES WITH EQUAL LEGS.

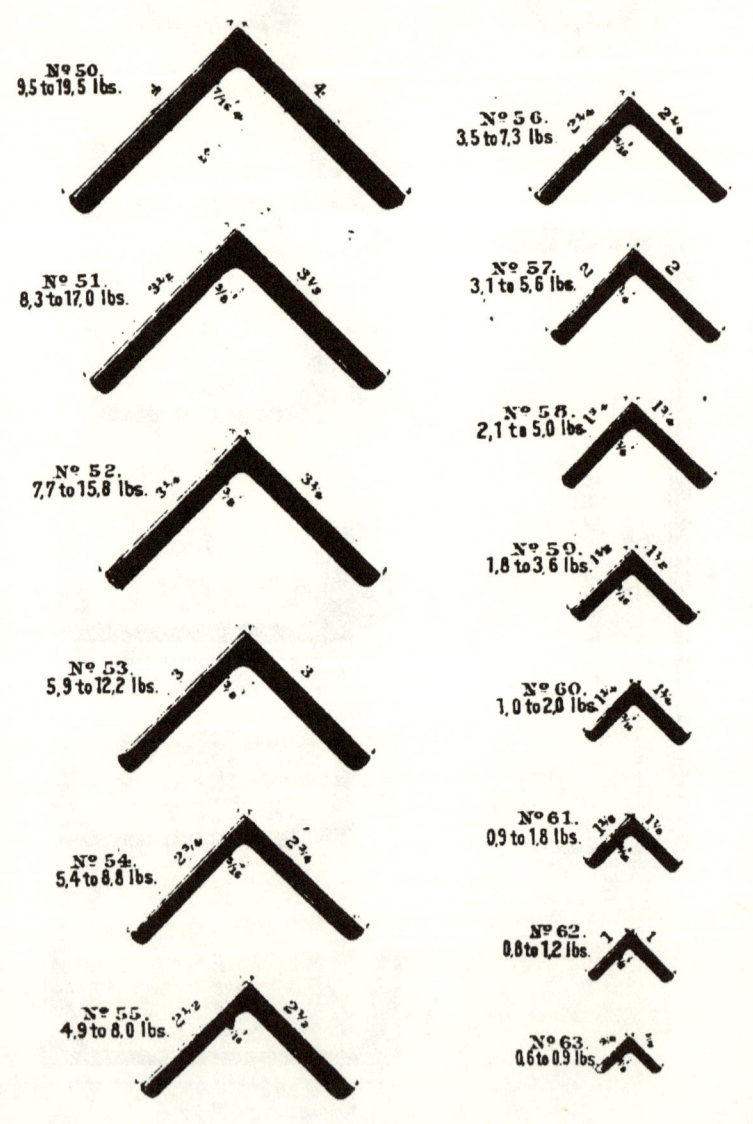

PITTSBURGH, PA.

ANGLES WITH UNEQUAL LEGS.

N° 65.
13,9 to 26,4 lbs.

N° 66.
10,8 to 22,0 lbs.

N° 67.
10,2 to 20,8 lbs.

N° 68.
9,5 to 19,5 lbs.

N° 69.
8,9 to 18,3 lbs.

N° 70.
8,3 to 17,0 lbs.

N° 71.
7,7 to 15,8 lbs.

N° 72.
4,2 to 8,5 lbs.

N° 73.
4,4 to 9,0 lbs.

N° 74.
4,0 to 8,1 lbs.

N° 75.
3,5 to 7,3 lbs.

N° 76.
2,6 to 4,0 lbs.

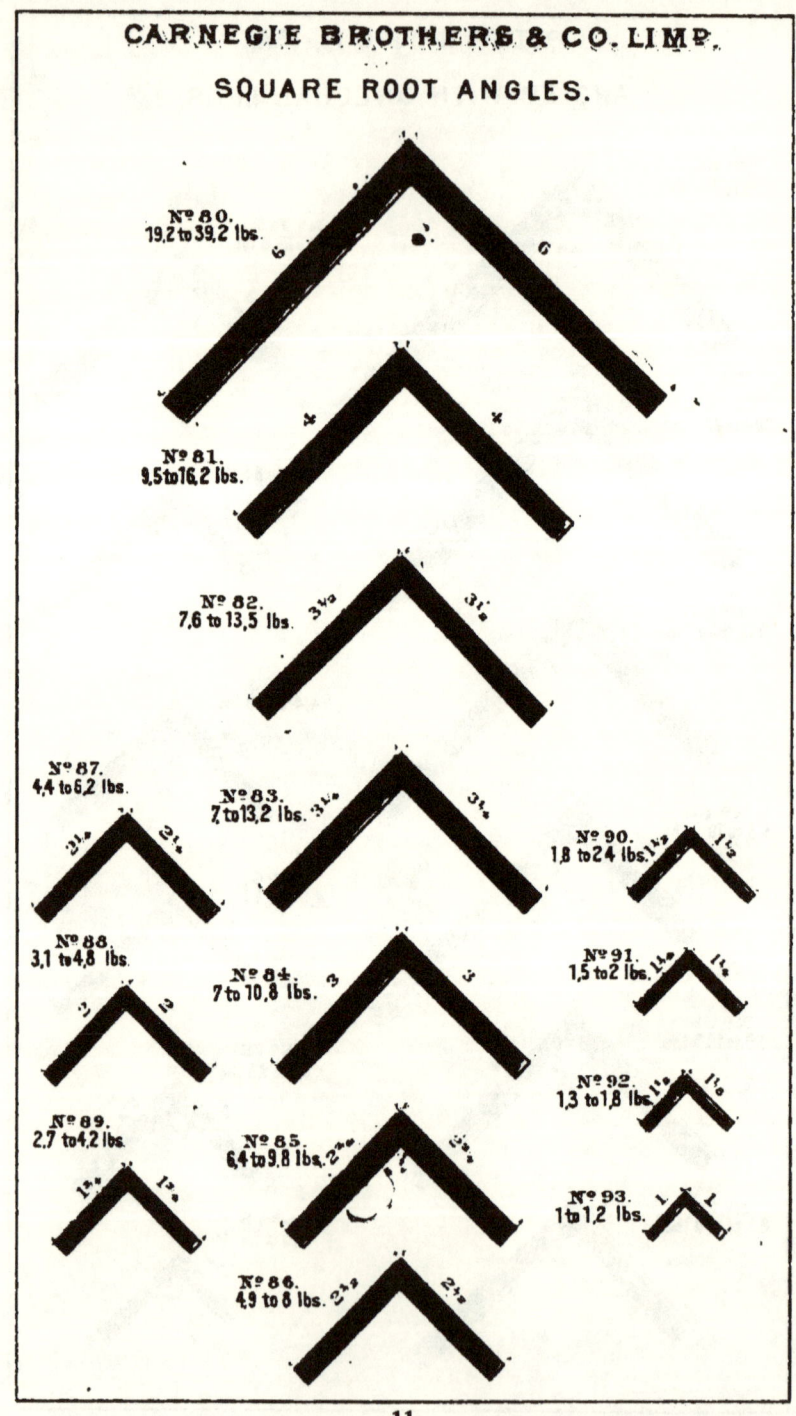

PITTSBURGH, PA.

COVER ANGLES.

Nº 95.
10.2 to 12.2 lbs.

Nº 96.
6.7 to 8.3 lbs.

OBTUSE ANGLE.

Nº 98.
6 lbs.
117°

STAR IRON.

Nº 100.
12 lbs.

Nº 101.
9½ lbs.

Nº 102.
7¼ lbs.

Nº 103.
5½ lbs.

Nº 104.
3¾ lbs.

Nº 105.
2.3 lbs.

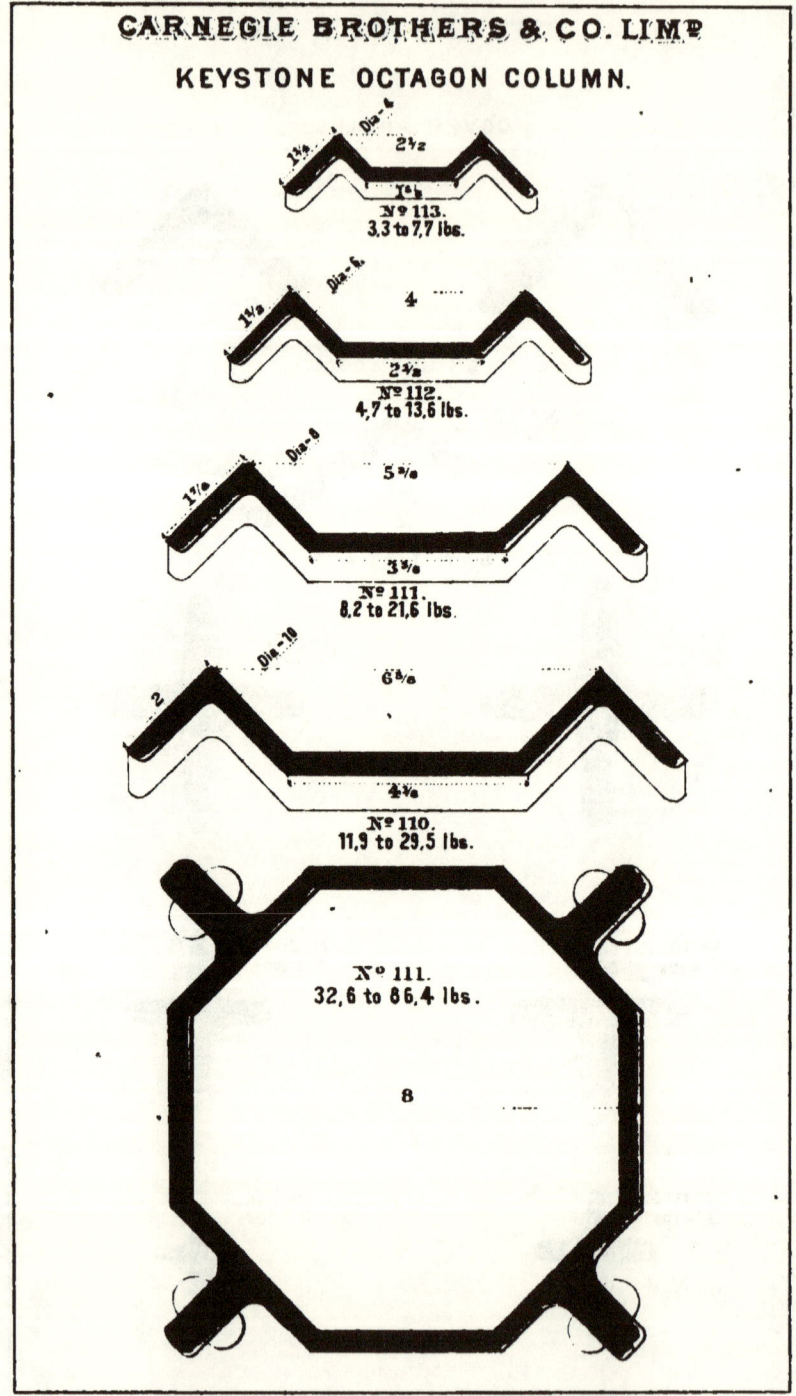

PITTSBURGH, PA.
PIPER'S PATENT RIVETLESS COLUMN.

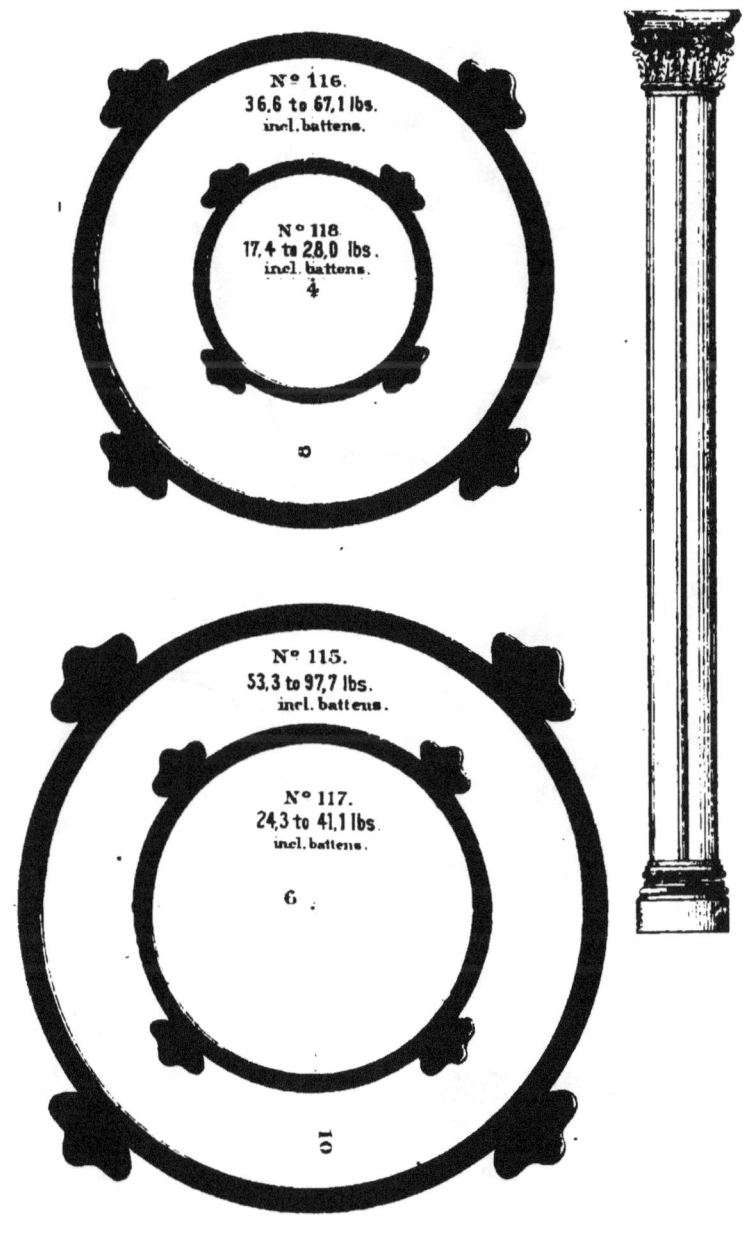

CARNEGIE BROTHERS & CO.
LIMITED.

CORRUGATED COLUMN.

N° 121.
42,6 to 87,9 lbs.

N° 120.
6,7 to 16,3 lbs.

N° 121.
5,3 to 11,0 lbs.

PATENT POST IRON.

HALF T.

N° 125.
6 to 8 lbs.

N° 126.
4,5 to 5,5 lbs.

N° 128.
4½ lbs.

N° 127.
4¾ lbs.

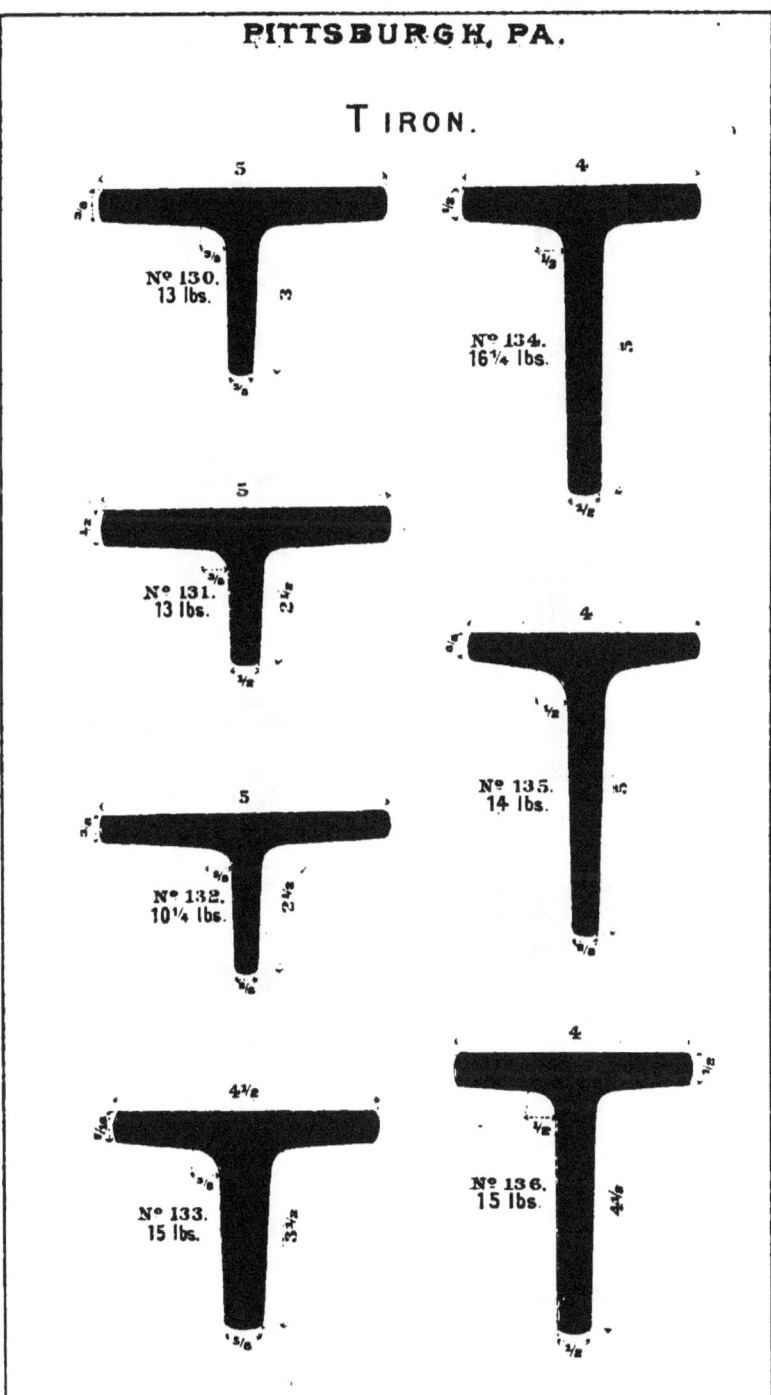

CARNEGIE BROTHERS & CO.
LIMITED.

Nº 137.
13½ lbs.

Nº 140.
9¼ lbs.

Nº 138.
14 lbs.

Nº 141.
8,8 lbs.

Nº 142.
7,5 lbs.

Nº 139.
12 lbs.

Nº 143.
8 lbs.

Nº 144.
6½ lbs.

PITTSBURGH, PA.

N° 145.
13, 8 lbs.

N° 146.
11¼ lbs.

N° 147.
11⅓ lbs.

N° 148.
10 lbs.

N° 149.
11¼ lbs.

N° 150.
9¼ lbs.

N° 151.
12¼ lbs.

N° 152.
11¾ lbs.

18

CARNEGIE BROTHERS & CO.
LIMITED.

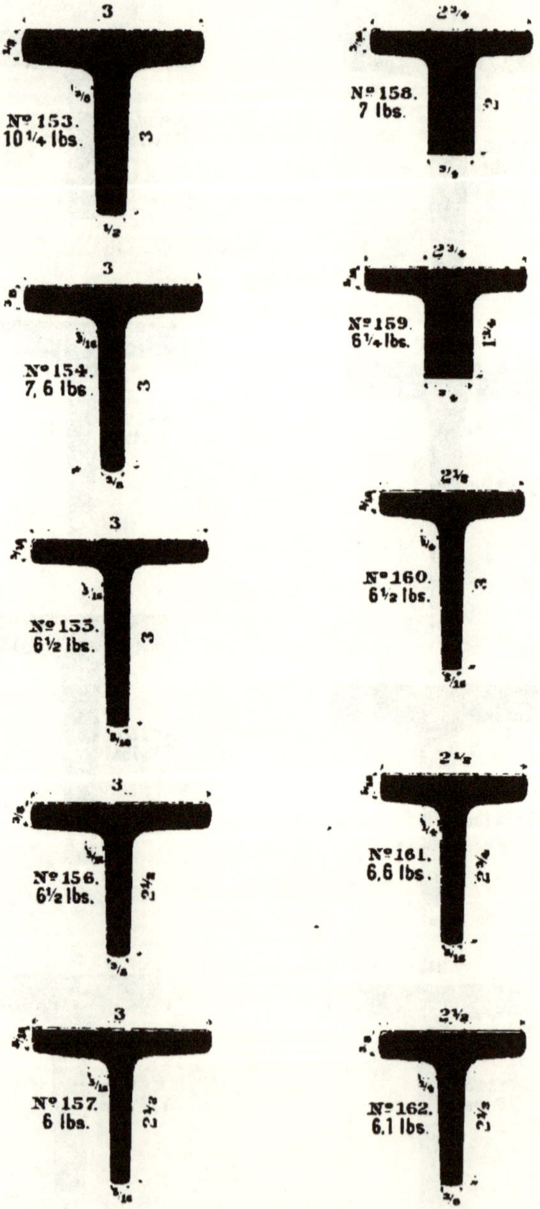

No. 153.
10¼ lbs.
3 × 3 × ⅜ × ½

No. 158.
7 lbs.
2¾ × ⅜ × ⅜

No. 154.
7.6 lbs.
3 × 3 × ⅜ × ⅜

No. 159.
6¼ lbs.
2¾ × ⅜ × 1¾

No. 155.
6½ lbs.
3 × 3 × ⅜ × ⅜

No. 160.
6½ lbs.
2⅞ × ¾ × 3

No. 156.
6½ lbs.
3 × 2½ × ⅜ × ⅜

No. 161.
6.6 lbs.
2½ × ⅜ × 2½

No. 157.
6 lbs.
3 × 2½ × ⅜ × ⅜

No. 162.
6.1 lbs.
2½ × ⅜ × 2½

PITTSBURGH, PA.

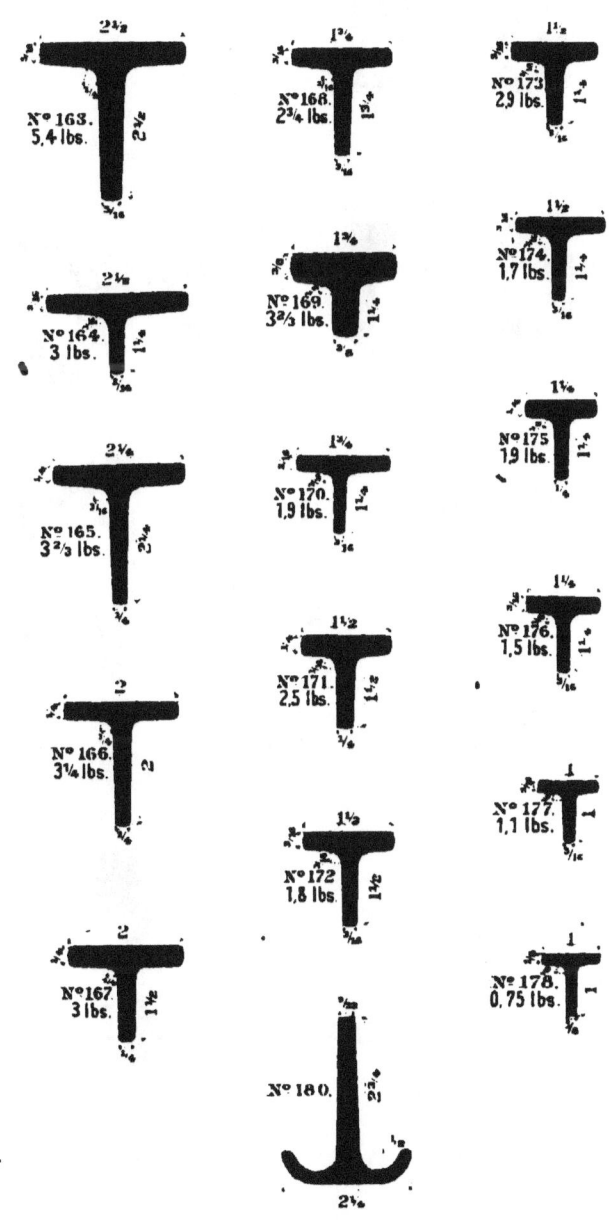

CARNEGIE BROTHERS & CO.
LIMITED.

GROOVED IRON.

No. 209. 0.42 lbs.
No. 208. 0.6 lbs.
No. 207. 0.6 lbs.

No. 206. 3/4 lbs.
No. 205. 1 lb.
No. 204. 1 1/4 lbs.
No. 203. 1 2/3 lbs.

No. 202. 3 lbs. R. 7/8.
No. 201. 2 1/2 lbs.
No. 200. 2 1/2 lbs.

HAND RAILS.

No. 195. 2.8 lbs.
No. 196. 3.1 lbs.

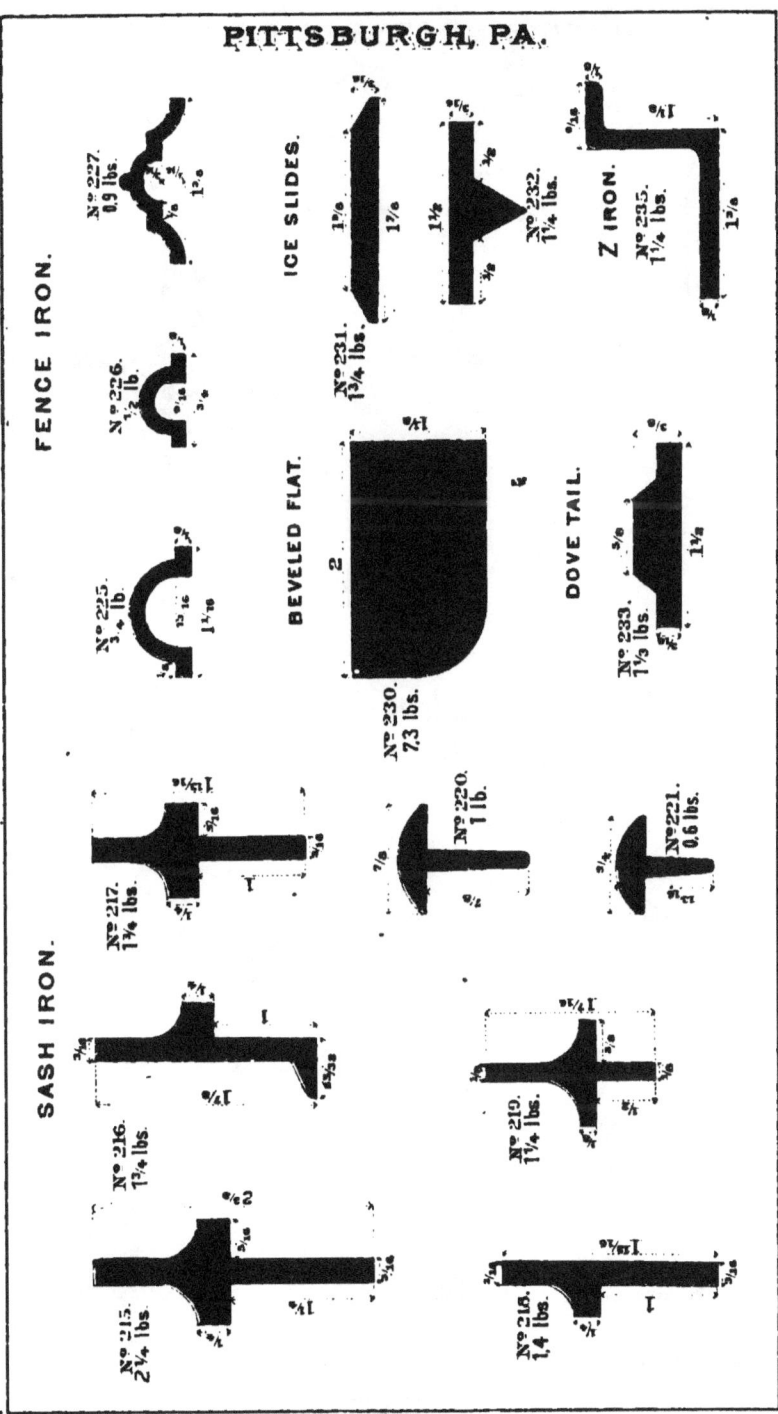

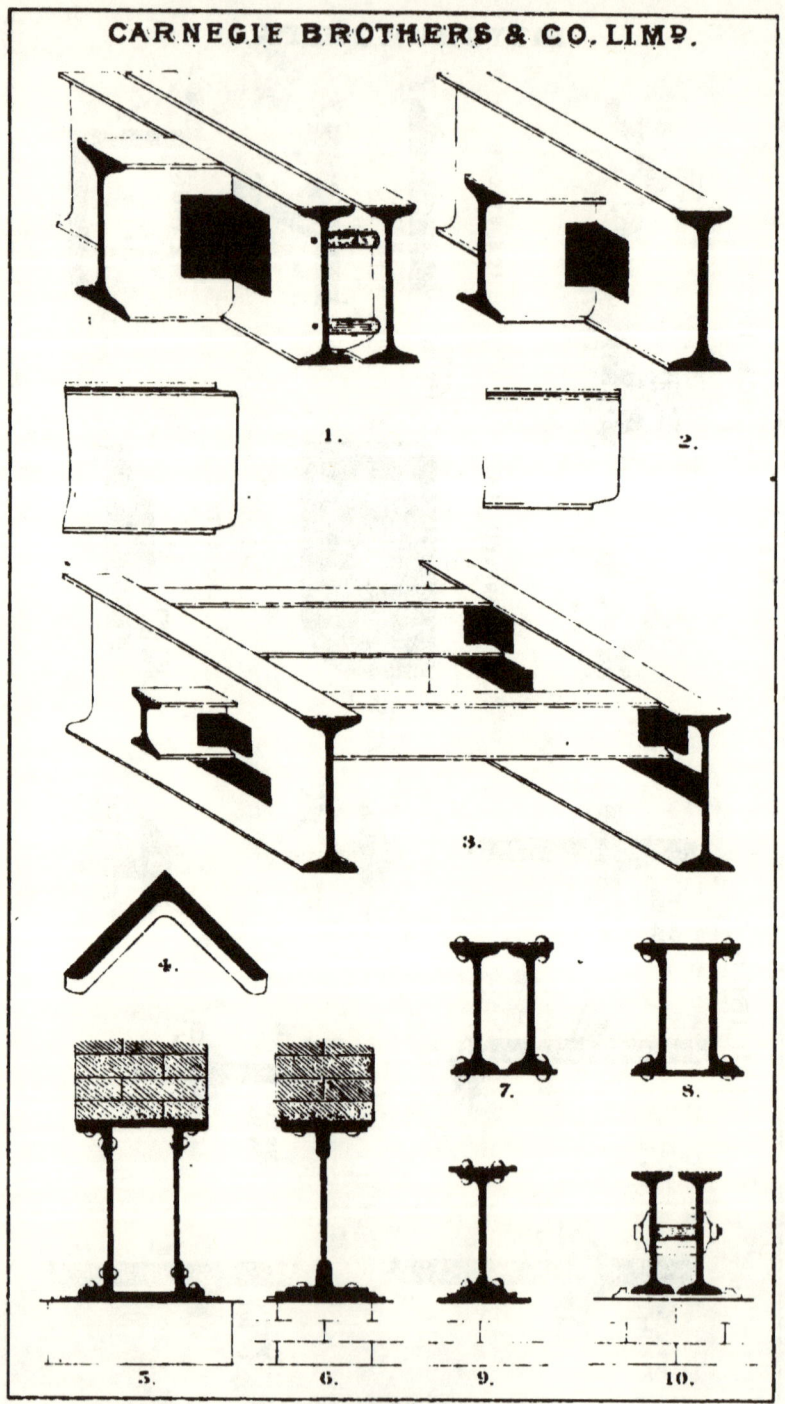

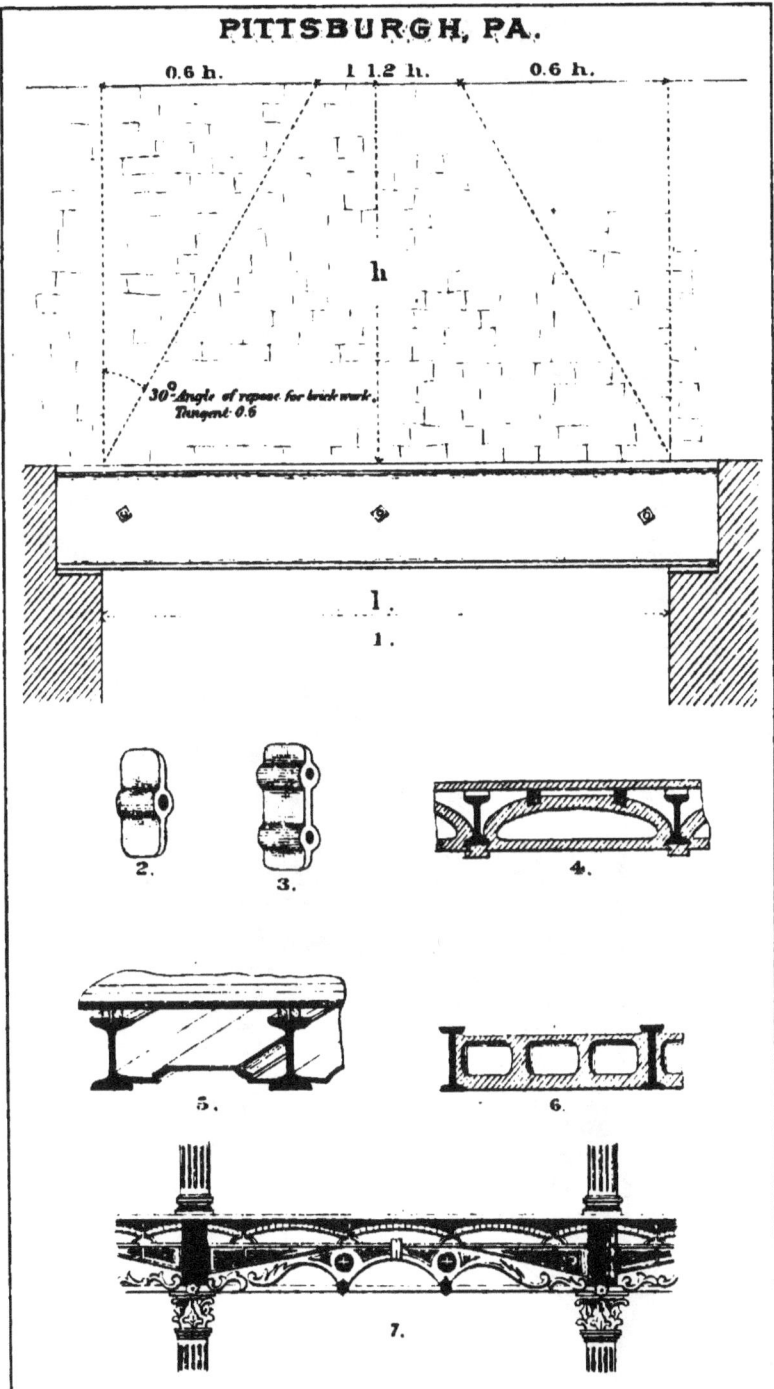

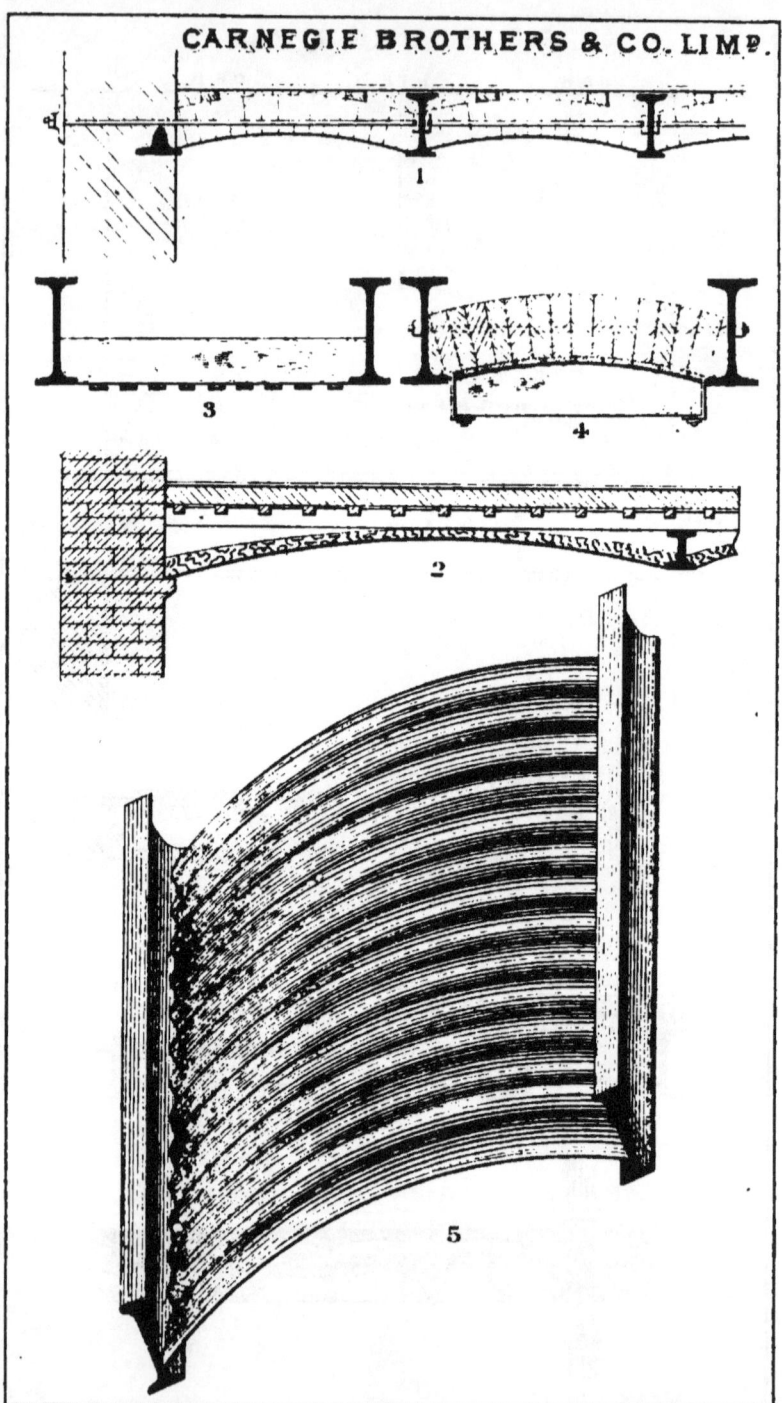

PITTSBURGH, PA.

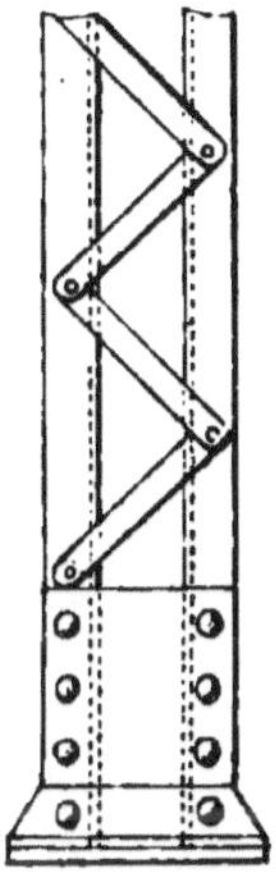

1.
2.  
3.
4. 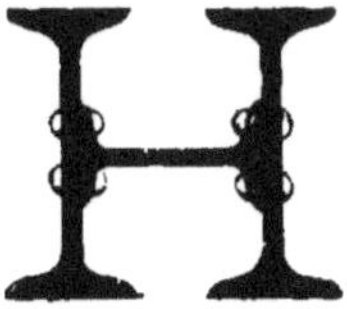
5.
6.

PRATT OR SINGLE QUADRANGULAR TRUSS.

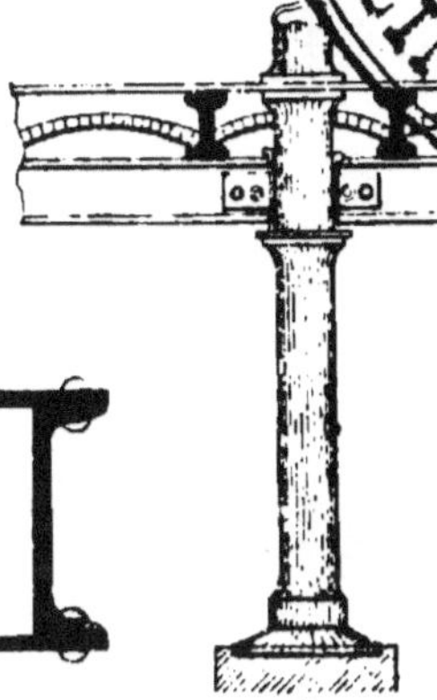

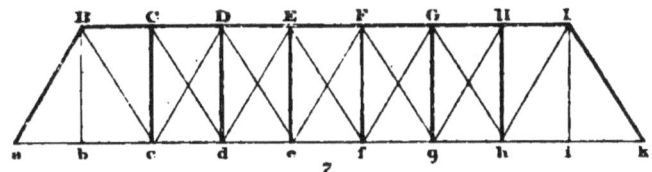

7.

WHIPPLE OR DOUBLE QUADRANGULAR TRUSS.

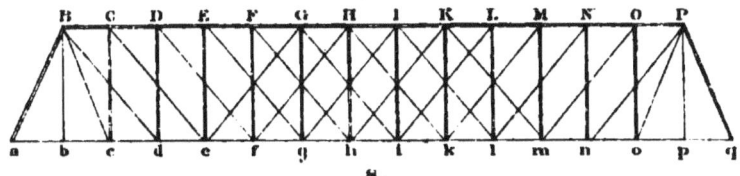

8.

CARNEGIE BROTHERS & CO.
LIMITED.
ADDITIONAL SHAPES.

PITTSBURGH, PA.

ADDITIONAL SHAPES.

CARNEGIE BROTHERS & CO.
LIMITED.
ADDITIONAL SHAPES.

PITTSBURGH, PA.

ADDITIONAL SHAPES.

CARNEGIE BROTHERS & CO.
LIMITED.
ADDITIONAL SHAPES.

EXPLANATION OF TABLES ON UNION IRON MILLS' EYEBEAMS.

Pages 33 to 55, Inclusive.

These tables are calculated for the lightest and heaviest weights to which each shape or size can be rolled, the term shape being meant to include the variable sections which are rolled in the same grooves by increasing or reducing the distance between the rolls. Each shape is designated by a single number.

These tables give:

I. In second column, the load which a beam will carry safely, distributed uniformly over its length, for the distances between supports, (or lengths of span,) given in first column;

II. In fifth to eleventh columns inclusive, the distances between centers at which beams should be placed in floors, to carry safely loads of 100, 125, 150, 175, 200, 250 and 300 lbs. per square foot (including the weight of the beams), for the distances between supports given in first column;

III. In third column, the deflection of the beams at center under these loads.

IV. In fourth column, the weight of the beam itself, for a length equal to the distance between supports.

To determine the load which a beam will carry exclusive of its own weight, the figures in fourth column must be subtracted from the figures in second column.

It is assumed in these tables that proper provision is made for preventing the compression flanges of the beams from deflecting sideways. They should be held in position at distances not exceeding twenty times the width of flange, otherwise the strain allowed should be reduced.

If the deflection of beams carrying plastered ceilings exceeds $\frac{1}{360}$th of the distance between supports, or $\frac{1}{30}$th of an inch per foot of this distance, there is danger of the ceiling cracking, as has been found by practical tests. This limit is indicated in the following tables by a cross line, beyond which the spans and loads must not be used for beams intended to carry plastered ceilings. It may generally be assumed, both for rolled and built beams, that the above limit is not exceeded so long as the depth of beam is not greater than $\frac{1}{24}$th of the distance between supports, or ½ inch per foot of this distance.

Inasmuch as the carrying capacity of beams increases largely with their depth, and it is therefore economical to use the greatest depth of beam consistent with the other conditions to which it is necessary to conform, (as clear hight, etc.,) the above cases of extreme deflection will rarely be met with in practice.

EXAMPLES OF APPLICATION OF TABLES.

I. What size and weight of beam 19′-6″ long in clear between walls, and therefore say 20′-0″ long between centers of supports, will be required to carry safely a uniformly distributed load of 15 tons, the weight of the beam included?

Answer: A 15″ beam, No. 1, heavy, 65 lbs. per foot, will be sufficient, since the safe load, as per table, for 20′ length,= 16.38 t.

It is evident, however, that a beam intermediate in weight between 50 lbs. and 65 lbs. can be used, to ascertain which, proceed as follows:

The safe load for a 15″ beam 50 lbs. per foot = 14.12 t. Since therefore an increase in the carrying capacity of beam, of 2.26 t., (16.38 t. — 14.12 t.,) requires an increase of its weight of 15 lbs., (65 lbs. — 50 lbs.,) therefore an increase of its carrying capacity of 0.88 t., (15 t. — 14.12 t.,) will require $\frac{0.88}{2.26} \times 15 = 6$ lbs. increase of weight of beam, *i. e.*, the beam should weigh 56 lbs. per foot.

II. A fire-proof floor 24′-6″ in clear between walls, weighing, inclusive of beams, 70 lbs. per square foot, (assumed,) is to be proportioned to carry an additional load of 130 lbs. per square foot; what size and weight of beams will be required, and how far apart should they be placed?

Answer: The total load = 200 lbs. per square foot, and the distance between supports = 25′, *i. e.* 6″ greater than the distance in clear between walls. By referring to tables, it will be seen that either light 12″ beams weighing 42 lbs. per foot, spaced 2.9 ft. between centers, or light 15″ beams, 50 lbs., spaced 5.8 ft. between centers, will answer the purpose, but since the 12″ beams for this span and load are beyond the cross-line, they must not be used, if intended to carry a plastered ceiling.

UNION IRON MILLS'
15-INCH EYEBEAM, No. 1, LIGHT,
50 LBS. PER FOOT.

Depth, 15″. Width of Flanges, 5.03″. Thickness of Web, 0.47″.
Maximum fiber strain = 12000 lbs. per square inch.

Distance between supports, in feet.	Safe load, uniformly distributed, (including weight of beam,) in tons of 2000 lbs.	Deflection under this load, in inches.	Weight of beam, in tons of 2000 lbs.	Proper distance, in feet, center to center of beams, for Safe Loads of						
				100 lbs. per sq. ft.	125 lbs. per sq. ft.	150 lbs. per sq. ft.	175 lbs. per sq. ft.	200 lbs. per sq. ft.	250 lbs. per sq. ft.	300 lbs. per sq. ft.
10	28.24	0.09	0.25	56.5	45.2	37.7	32.3	28.2	22.6	18.8
11	25.67	0.11	0.28	46.7	37.4	31.1	26.7	23.3	18.7	15.6
12	23.53	0.13	0.30	39.2	31.4	26.1	22.4	19.6	15.7	13.1
13	21.72	0.16	0.33	33.4	26.7	22.3	19.1	16.7	13.4	11.1
14	20.17	0.18	0.35	28.8	23.0	19.2	16.5	14.4	11.5	9.6
15	18.83	0.21	0.38	25.1	20.1	16.7	14.3	12.6	10.0	8.4
16	17.65	0.24	0.40	22.1	17.7	14.7	12.6	11.0	8.8	7.4
17	16.61	0.27	0.43	19.5	15.6	13.0	11.1	9.8	7.8	6.5
18	15.69	0.30	0.45	17.4	13.9	11.6	9.9	8.7	7.0	5.8
19	14.86	0.33	0.48	15.6	12.5	10.4	8.9	7.8	6.2	5.2
20	14.12	0.37	0.50	14.1	11.3	9.4	8.1	7.1	5.6	4.7
21	13.45	0.41	0.53	12.8	10.2	8.5	7.3	6.4	5.1	4.3
22	12.84	0.45	0.55	11.7	9.3	7.8	6.7	5.8	4.7	3.9
23	12.28	0.49	0.58	10.7	8.6	7.1	6.1	5.3	4.3	3.6
24	11.77	0.53	0.60	9.8	7.8	6.5	5.6	4.9	3.9	3.3
25	11.30	0.58	0.63	9.0	7.2	6.0	5.1	4.5	3.6	3.0
26	10.86	0.62	0.65	8.4	6.7	5.6	4.8	4.2	3.4	2.8
27	10.46	0.67	0.68	7.7	6.2	5.1	4.4	3.9	3.1	2.6
28	10.09	0.72	0.70	7.2	5.8	4.8	4.1	3.6	2.9	2.4
29	9.74	0.78	0.73	6.7	5.4	4.5	3.8	3.4	2.7	2.2
30	9.41	0.83	0.75	6.3	5.0	4.2	3.6	3.1	2.5	2.1
31	9.11	0.89	0.78	5.9	4.7	3.9	3.4	2.9	2.4	2.0
32	8.83	0.94	0.80	5.5	4.4	3.7	3.2	2.8	2.2	1.8
33	8.56	1.00	0.83	5.2	4.2	3.5	3.0	2.6	2.1	1.7
34	8.31	1.07	0.85	4.9	3.9	3.3	2.8	2.4	2.0	1.6
35	8.07	1.13	0.88	4.6	3.7	3.1	2.6	2.3	1.8	1.5
36	7.84	1.19	0.90	4.4	3.5	2.9	2.5	2.2	1.7	1.5
37	7.63	1.26	0.93	4.1	3.3	2.7	2.4	2.1	1.6	1.4
38	7.43	1.33	0.95	3.9	3.1	2.6	2.2	2.0	1.6	1.3
39	7.24	1.40	0.98	3.7	3.0	2.5	2.1	1.9	1.5	1.2

UNION IRON MILLS'

15-INCH EYEBEAM, No. 1, HEAVY, 65 LBS. PER FOOT.

Depth, 15″. Width of Flanges, 5.33″. Thickness of Web, 0.77″.
Maximum fiber strain = 12000 lbs. per square inch.

Distance between supports, in feet.	Safe load, uniformly distributed, (including weight of beam,) in tons of 2000 lbs.	Deflection under this load, in inches.	Weight of beam, in tons of 2000 lbs.	Proper distance, in feet, center to center of beams, for Safe Loads of							
				100 lbs. per sq. ft.	125 lbs. per sq. ft.	150 lbs. per sq. ft.	175 lbs. per sq. ft.	200 lbs. per sq. ft.	250 lbs. per sq. ft.	300 lbs. per sq. ft.
10	32.76	0.09	0.33	65.5	52.4	43.7	37.4	32.8	26.2	21.8
11	29.78	0.11	0.36	54.1	43.3	36.1	30.9	27.1	21.7	18.0
12	27.30	0.13	0.39	45.5	36.4	30.3	26.0	22.8	18.2	15.2
13	25.20	0.16	0.42	33.8	31.0	25.8	22.2	19.4	15.5	12.9
14	23.40	0.18	0.46	33.4	26.7	22.3	19.1	16.7	13.4	11.1
15	21.84	0.21	0.49	29.1	23.3	19.4	16.6	14.6	11.6	9.7
16	20.48	0.24	0.52	25.6	20.5	17.1	14.6	12.8	10.2	8.5
17	19.27	0.27	0.55	22.7	18.1	15.1	13.0	11.3	9.1	7.6
18	18.20	0.30	0.59	20.2	16.2	13.5	11.6	10.1	8.1	6.7
19	17.24	0.33	0.62	18.1	14.5	12.1	10.4	9.1	7.3	6.0
20	16.38	0.37	0.65	16.4	13.1	10.9	9.4	8.2	6.6	5.5
21	15.60	0.41	0.68	14.9	11.9	9.9	8.5	7.4	5.9	5.0
22	14.89	0.45	0.72	13.5	10.8	9.0	7.7	6.8	5.4	4.5
23	14.24	0.49	0.75	12.4	9.9	8.3	7.1	6.2	5.0	4.1
24	13.65	0.53	0.78	11.4	9.1	7.6	6.5	5.7	4.6	3.8
25	13.10	0.58	0.81	10.5	8.4	7.0	6.0	5.2	4.2	3.5
26	12.60	0.62	0.85	9.7	7.8	6.5	5.5	4.8	3.9	3.2
27	12.13	0.67	0.88	9.0	7.2	6.0	5.1	4.5	3.6	3.0
28	11.70	0.72	0.91	8.4	6.7	5.6	4.8	4.2	3.3	2.8
29	11.30	0.78	0.94	7.8	6.2	5.2	4.4	3.9	3.1	2.6
30	10.92	0.83	0.98	7.3	5.8	4.9	4.2	3.6	2.9	2.4
31	10.57	0.89	1.01	6.8	5.5	4.5	3.9	3.4	2.7	2.3
32	10.24	0.95	1.04	6.4	5.1	4.3	3.7	3.2	2.6	2.1
33	9.93	1.01	1.07	6.0	4.8	4.0	3.4	3.0	2.4	2.0
34	9.64	1.07	1.11	5.7	4.5	3.8	3.2	2.8	2.3	1.9
35	9.36	1.13	1.14	5.3	4.3	3.6	3.1	2.7	2.1	1.8
36	9.10	1.20	1.17	5.1	4.0	3.4	2.9	2.5	2.0	1.7
37	8.85	1.26	1.20	4.8	3.8	3.2	2.7	2.4	1.9	1.6
38	8.62	1.33	1.24	4.5	3.6	3.0	2.6	2.3	1.8	1.5
39	8.40	1.40	1.27	4.3	3.4	2.9	2.5	2.2	1.7	1.4

UNION IRON MILLS'

15-INCH EYEBEAM, No. 2, LIGHT, 67 LBS. PER FOOT.

Depth, 15″. Width of Flanges, 5.55″. Thickness of Web, 0.67″.
Maximum fiber strain = 12000 lbs. per square inch.

Distance between supports, in feet.	Safe load, uniformly distributed, (including weight of beam,) in tons of 2000 lbs.	Deflection under this load, in inches.	Weight of beam, in tons of 2000 lbs.	Proper distance, in feet, center to center of beams, for Safe Loads of						
				100 lbs. per sq. ft.	125 lbs. per sq. ft.	150 lbs. per sq. ft.	175 lbs. per sq. ft.	200 lbs. per sq. ft.	250 lbs. per sq. ft.	300 lbs. per sq. ft.
10	36.12	0.09	0.34	72.2	57.8	48.2	41.3	36.1	28.9	24.1
11	32.84	0.11	0.37	59.7	47.8	39.8	34.1	29.9	23.9	19.9
12	30.10	0.13	0.40	50.2	40.1	33.4	28.7	25.1	20.1	16.7
13	27.78	0.16	0.44	42.7	34.2	28.5	24.4	21.4	17.1	14.2
14	25.80	0.18	0.47	36.9	29.5	24.6	21.1	18.4	14.7	12.3
15	24.08	0.21	0.50	32.1	25.7	21.4	18.3	16.1	12.8	10.7
16	22.58	0.24	0.54	28.2	22.6	18.8	16.1	14.1	11.3	9.4
17	21.25	0.27	0.57	25.0	20.0	16.7	14.3	12.5	10.0	8.3
18	20.07	0.30	0.60	22.3	17.8	14.9	12.7	11.2	8.9	7.4
19	19.01	0.33	0.64	20.0	16.0	13.3	11.4	10.0	8.0	6.7
20	18.06	0.37	0.67	18.1	14.4	12.0	10.3	9.0	7.2	6.0
21	17.20	0.41	0.70	16.4	13.1	10.9	9.4	8.2	6.6	5.5
22	16.42	0.45	0.74	14.9	11.9	10.0	8.5	7.5	6.0	5.0
23	15.70	0.49	0.77	13.7	10.9	9.1	7.8	6.8	5.5	4.6
24	15.05	0.53	0.80	12.5	10.0	8.4	7.2	6.3	5.0	4.2
25	14.45	0.58	0.84	11.6	9.2	7.7	6.7	5.8	4.6	3.9
26	13.89	0.62	0.87	10.7	8.5	7.1	6.1	5.3	4.3	3.6
27	13.38	0.67	0.91	9.9	7.9	6.6	5.6	5.0	4.0	3.3
28	12.90	0.72	0.94	9.2	7.4	6.2	5.3	4.6	3.7	3.1
29	12.46	0.78	0.97	8.6	6.9	5.7	4.9	4.3	3.4	2.9
30	12.04	0.83	1.01	8.0	6.4	5.4	4.6	4.0	3.2	2.7
31	11.65	0.89	1.04	7.5	6.0	5.0	4.3	3.8	3.0	2.5
32	11.29	0.95	1.07	7.1	5.6	4.7	4.0	3.5	2.8	2.4
33	10.95	1.01	1.11	6.6	5.3	4.4	3.8	3.3	2.7	2.2
34	10.62	1.07	1.14	6.2	5.0	4.1	3.6	3.1	2.5	2.1
35	10.32	1.13	1.17	5.9	4.7	3.9	3.4	2.9	2.4	2.0
36	10.03	1.20	1.21	5.6	4.5	3.7	3.2	2.8	2.2	1.9
37	9.76	1.26	1.24	5.3	4.2	3.5	3.0	2.6	2.1	1.8
38	9.51	1.33	1.27	5.0	4.0	3.3	2.9	2.5	2.0	1.7
39	9.26	1.40	1.31	4.7	3.8	3.2	2.7	2.4	1.9	1.6

UNION IRON MILLS'

15-INCH EYEBEAM, No. 2, HEAVY, 80 LBS. PER FOOT.

Depth, 15″. Width of Flanges, 5.81″. Thickness of Web, 0.93″.
Maximum fiber strain = 12000 lbs. per square inch.

Distance between supports, in feet	Safe load, uniformly distributed, (including weight of beam,) in tons of 2000 lbs.	Deflection under this load, in inches	Weight of beam, in tons of 2000 lbs.	Proper distance, in feet, center to center. of beams, for Safe Loads of						
				100 lbs. per sq. ft.	125 lbs. per sq. ft.	150 lbs. per sq. ft.	175 lbs. per sq. ft.	200 lbs. per sq. ft.	250 lbs. per sq. ft.	300 lbs. per sq. ft.
10	40.00	0.09	0.40	80.0	64.0	53.3	45.7	40.0	32.0	26.7
11	36.36	0.11	0.44	66.1	52.9	44.1	37.8	33.1	26.4	22.0
12	33.33	0.13	0.48	55.6	44.4	37.0	31.7	27.8	22.2	18.5
13	30.77	0.16	0.52	47.3	37.9	31.6	27.1	23.7	18.9	15.8
14	28.57	0.18	0.56	40.8	32.6	27.2	23.3	20.4	16.3	13.6
15	26.67	0.21	0.60	35.6	28.4	23.7	20.3	17.8	14.2	11.9
16	25.00	0.24	0.64	31.3	25.0	20.8	17.9	15.6	12.5	10.4
17	23.53	0.27	0.68	27.7	22.1	18.5	15.8	13.8	11.1	9.2
18	22.22	0.30	0.72	24.7	19.8	16.5	14.1	12.3	9.9	8.2
19	21.05	0.33	0.76	22.2	17.7	14.8	12.6	11.1	8.9	7.4
20	20.00	0.37	0.80	20.0	16.0	13.3	11.4	10.0	8.0	6.7
21	19.05	0.41	0.84	18.1	14.5	12.1	10.4	9.1	7.3	6.0
22	18.18	0.45	0.88	16.5	13.2	11.0	9.4	8.3	6.6	5.5
23	17.39	0.49	0.92	15.1	12.1	10.1	8.6	7.6	6.0	5.0
24	16.67	0.53	0.96	13.9	11.1	9.3	7.9	6.9	5.6	4.6
25	16.00	0.58	1.00	12.8	10.2	8.5	7.3	6.4	5.1	4.3
26	15.38	0.62	1.04	11.8	9.5	7.9	6.8	5.9	4.7	3.9
27	14.81	0.67	1.08	11.0	8.8	7.3	6.3	5.5	4.4	3.7
28	14.29	0.72	1.12	10.2	8.2	6.8	5.8	5.1	4.1	3.4
29	13.79	0.78	1.16	9.5	7.6	6.3	5.4	4.8	3.8	3.2
30	13.33	0.83	1.20	8.9	7.1	5.9	5.1	4.4	3.6	3.0
31	12.90	0.89	1.24	8.3	6.6	5.5	4.8	4.2	3.3	2.8
32	12.50	0.95	1.28	7.8	6.2	5.2	4.5	3.9	3.1	2.6
33	12.12	1.01	1.32	7.3	5.9	4.9	4.2	3.7	2.9	2.4
34	11.76	1.07	1.36	6.9	5.5	4.6	3.9	3.5	2.8	2.3
35	11.43	1.13	1.40	6.5	5.2	4.3	3.7	3.3	2.6	2.2
36	11.11	1.20	1.44	6.2	4.9	4.1	3.5	3.1	2.5	2.1
37	10.81	1.26	1.48	5.8	4.7	3.9	3.3	2.9	2.3	1.9
38	10.53	1.33	1.52	5.5	4.4	3.7	3.2	2.8	2.2	1.8
39	10.26	1.40	1.56	5.3	4.2	3.5	3.0	2.6	2.1	1.8

UNION IRON MILLS'

12-INCH EYEBEAM, No. 3, LIGHT, 42 LBS. PER FOOT.

Depth, 12″. Width of Flanges, 4.64″. Thickness of Web, 0.51″.
Maximum fiber strain = 12000 lbs. per square inch.

Distance between supports, in feet	Safe load, uniformly distributed, (including weight of beam,) in tons of 2000 lbs.	Deflection under this load, in inches.	Weight of beam, in tons of 2000 lbs.	Proper distance, in feet, center to center of beams, for Safe Loads of						
				100 lbs. per sq. ft.	125 lbs. per sq. ft.	150 lbs. per sq. ft.	175 lbs. per sq. ft.	200 lbs. per sq. ft.	250 lbs. per sq. ft.	300 lbs. per sq. ft.
10	18.36	0.12	0.21	36.7	29.4	24.5	21.0	18.4	14.7	12.2
11	16.69	0.14	0.23	30.3	24.3	20.2	17.3	15.2	12.1	10.1
12	15.30	0.17	0.25	25.5	20.4	17.0	14.6	12.8	10.2	8.5
13	14.12	0.20	0.27	21.7	17.4	14.5	12.4	10.9	8.7	7.2
14	13.11	0.23	0.29	18.7	15.0	12.5	10.7	9.4	7.5	6.2
15	12.24	0.26	0.32	16.3	13.1	10.9	9.3	8.2	6.5	5.4
16	11.48	0.30	0.34	14.4	11.5	9.6	8.2	7.2	5.7	4.8
17	10.80	0.33	0.36	12.7	10.2	8.5	7.3	6.4	5.1	4.2
18	10.20	0.37	0.38	11.3	9.1	7.6	6.5	5.7	4.5	3.8
19	9.66	0.42	0.40	10.2	8.1	6.8	5.8	5.1	4.1	3.4
20	9.18	0.46	0.42	9.2	7.3	6.1	5.2	4.6	3.7	3.1
21	8.74	0.51	0.44	8.3	6.7	5.5	4.8	4.2	3.3	2.8
22	8.35	0.56	0.46	7.6	6.1	5.0	4.3	3.8	3.0	2.5
23	7.98	0.61	0.48	6.9	5.6	4.6	4.0	3.5	2.8	2.3
24	7.65	0.66	0.50	6.4	5.1	4.2	3.6	3.2	2.6	2.1
25	7.34	0.72	0.53	5.9	4.7	3.9	3.3	2.9	2.4	2.0
26	7.06	0.78	0.55	5.4	4.3	3.6	3.1	2.7	2.2	1.8
27	6.80	0.84	0.57	5.0	4.0	3.3	2.9	2.5	2.0	1.7
28	6.56	0.90	0.59	4.7	3.7	3.1	2.7	2.3	1.9	1.6
29	6.33	0.97	0.61	4.4	3.5	2.9	2.5	2.2	1.7	1.5
30	6.12	1.04	0.63	4.1	3.3	2.7	2.3	2.0	1.6	1.4
31	5.92	1.11	0.65	3.8	3.1	2.5	2.2	1.9	1.5	1.3
32	5.74	1.18	0.67	3.6	2.9	2.3	2.0	1.8	1.4	1.2
33	5.56	1.26	0.69	3.4	2.7	2.2	1.9	1.7	1.3	1.1
34	5.40	1.34	0.71	3.2	2.5	2.1	1.8	1.6	1.3	1.1
35	5.25	1.42	0.74	3.0	2.4	2.0	1.7	1.5	1.2	1.0
36	5.10	1.50	0.76	2.8	2.2	1.9	1.6	1.4	1.1	0.9
37	4.96	1.58	0.78	2.6	2.1	1.8	1.5	1.3	1.1	0.9
38	4.83	1.67	0.80	2.5	2.0	1.7	1.5	1.3	1.0	0.8
39	4.71	1.76	0.82	2.4	1.9	1.6	1.4	1.2	1.0	0.8

UNION IRON MILLS'

12-INCH EYEBEAM, No. 3, HEAVY, 60 LBS. PER FOOT.

Depth, 12″. Width of Flanges, 5.09″. Thickness of Web, 0.96″.
Maximum fiber strain = 12000 lbs. per square inch.

Distance between supports, in feet	Safe load, uniformly distributed, (including weight of beam,) in tons of 2000 lbs.	Deflection under this load, in inches	Weight of beam, in tons of 2000 lbs.	Proper distance, in feet, center to center of beams, for Safe Loads of						
				100 lbs. per sq. ft.	125 lbs. per sq. ft.	150 lbs. per sq. ft.	175 lbs. per sq. ft.	200 lbs. per sq. ft.	250 lbs. per sq. ft.	300 lbs. per sq. ft.
10	22.68	0.12	0.30	45.4	36.3	30.2	25.9	22.7	18.1	15.1
11	20.62	0.14	0.33	37.5	30.0	25.0	21.4	18.7	15.0	12.5
12	18.90	0.17	0.36	31.5	25.2	21.0	18.0	15.8	12.6	10.5
13	17.45	0.20	0.39	26.8	21.5	17.9	15.3	13.4	10.7	8.9
14	16.20	0.23	0.42	23.1	18.5	15.4	13.2	11.6	9.3	7.7
15	15.12	0.26	0.45	20.2	16.1	13.4	11.5	10.1	8.1	6.7
16	14.18	0.30	0.48	17.7	14.2	11.8	10.1	8.9	7.1	5.9
17	13.34	0.33	0.51	15.7	12.6	10.5	9.0	7.8	6.3	5.2
18	12.60	0.37	0.54	14.0	11.2	9.3	8.0	7.0	5.6	4.7
19	11.94	0.42	0.57	12.6	10.1	8.4	7.2	6.3	5.0	4.2
20	11.34	0.46	0.60	11.3	9.1	7.6	6.5	5.7	4.5	3.8
21	10.80	0.51	0.63	10.3	8.2	6.9	5.9	5.2	4.1	3.4
22	10.31	0.56	0.66	9.4	7.5	6.2	5.4	4.7	3.7	3.1
23	9.86	0.61	0.69	8.6	6.9	5.7	4.9	4.3	3.4	2.9
24	9.45	0.66	0.72	7.9	6.3	5.3	4.5	3.9	3.1	2.6
25	9.07	0.72	0.75	7.3	5.8	4.9	4.2	3.6	2.9	2.4
26	8.72	0.78	0.78	6.7	5.4	4.5	3.9	3.3	2.7	2.2
27	8.40	0.84	0.81	6.2	5.0	4.2	3.6	3.1	2.5	2.1
28	8.10	0.90	0.84	5.8	4.6	3.9	3.3	2.9	2.3	1.9
29	7.82	0.97	0.87	5.4	4.3	3.6	3.1	2.7	2.1	1.8
30	7.56	1.04	0.90	5.0	4.0	3.4	2.9	2.5	2.0	1.7
31	7.32	1.11	0.93	4.7	3.8	3.2	2.7	2.4	1.9	1.6
32	7.09	1.18	0.96	4.4	3.5	3.0	2.5	2.2	1.8	1.5
33	6.87	1.26	0.99	4.2	3.3	2.8	2.4	2.1	1.7	1.4
34	6.67	1.34	1.02	3.9	3.1	2.6	2.2	2.0	1.6	1.3
35	6.48	1.42	1.05	3.7	3.0	2.5	2.1	1.9	1.5	1.2
36	6.30	1.50	1.08	3.5	2.8	2.3	2.0	1.8	1.4	1.2
37	6.13	1.58	1.11	3.3	2.6	2.2	1.9	1.7	1.3	1.1
38	5.97	1.67	1.14	3.1	2.5	2.1	1.8	1.6	1.3	1.0
39	5.82	1.76	1.17	3.0	2.4	2.0	1.7	1.5	1.2	1.0

UNION IRON MILLS'
10½-INCH EYEBEAM, No. 4, LIGHT, 31½ LBS. PER FOOT.

Depth, 10½″. Width of Flanges, 4.54″. Thickness of Web, 0.41″.
Maximum fiber strain = 12000 lbs. per square inch.

Distance between supports, in feet.	Safe load, uniformly distributed, (including weight of beam,) in tons of 2000 lbs.	Deflection under this load, in inches.	Weight of beam, in tons of 2000 lbs.	Proper distance, in feet, center to center of beams, for Safe Loads of							
				100 lbs. per sq. ft.	125 lbs. per sq. ft.	150 lbs. per sq. ft.	175 lbs. per sq. ft.	200 lbs. per sq. ft.	250 lbs. per sq. ft.	300 lbs. per sq. ft.
10	12.56	0.13	0.16	25.1	20.1	16.7	14.4	12.6	10.0	8.4
11	11.42	0.16	0.17	20.8	16.6	13.8	11.9	10.4	8.3	6.9
12	10.47	0.19	0.19	17.5	14.0	11.6	10.0	8.7	7.0	5.8
13	9.66	0.22	0.21	14.9	11.9	9.9	8.5	7.4	5.9	5.0
14	8.97	0.26	0.22	12.8	10.2	8.5	7.3	6.4	5.1	4.3
15	8.37	0.30	0.24	11.2	8.9	7.4	6.4	5.6	4.5	3.7
16	7.85	0.34	0.25	9.8	7.8	6.5	5.6	4.9	3.9	3.3
17	7.39	0.38	0.27	8.7	7.0	5.8	5.0	4.3	3.5	2.9
18	6.98	0.43	0.28	7.8	6.2	5.2	4.4	3.9	3.1	2.6
19	6.61	0.48	0.30	7.0	5.6	4.6	4.0	3.5	2.8	2.3
20	6.28	0.53	0.32	6.3	5.0	4.2	3.6	3.1	2.5	2.1
21	5.98	0.58	0.33	5.7	4.6	3.8	3.3	2.8	2.3	1.9
22	5.71	0.64	0.35	5.2	4.2	3.5	3.0	2.6	2.1	1.7
23	5.46	0.70	0.36	4.8	3.8	3.2	2.7	2.4	1.9	1.6
24	5.23	0.76	0.38	4.4	3.5	2.9	2.5	2.2	1.7	1.5
25	5.02	0.82	0.39	4.0	3.2	2.7	2.3	2.0	1.6	1.3
26	4.83	0.89	0.41	3.7	3.0	2.5	2.1	1.9	1.5	1.2
27	4.65	0.96	0.43	3.4	2.8	2.3	2.0	1.7	1.4	1.1
28	4.49	1.03	0.44	3.2	2.6	2.1	1.8	1.6	1.3	1.1
29	4.33	1.11	0.46	3.0	2.4	2.0	1.7	1.5	1.2	1.0
30	4.19	1.19	0.47	2.8	2.2	1.9	1.6	1.4	1.1	.9
31	4.05	1.27	0.49	2.6	2.1	1.7	1.5	1.3	1.0	.9
32	3.93	1.35	0.50	2.5	2.0	1.6	1.4	1.2	1.0	.8
33	3.81	1.44	0.52	2.3	1.8	1.5	1.3	1.2	.9	.8
34	3.69	1.53	0.54	2.2	1.7	1.4	1.2	1.1	.9	.7
35	3.59	1.62	0.55	2.1	1.6	1.4	1.2	1.0	.8	.7
36	3.49	1.71	0.57	1.9	1.6	1.3	1.1	1.0	.8	.6
37	3.39	1.80	0.58	1.8	1.5	1.2	1.1	.9	.7	.6
38	3.31	1.90	0.60	1.7	1.4	1.2	1.0	.9	.7	.6
39	3.22	2.01	0.61	1.7	1.3	1.1	.9	.8	.7	.6

UNION IRON MILLS'
10½-INCH EYEBEAM, No. 4, HEAVY, 45 LBS. PER FOOT.

Depth, 10½". Width of Flanges, 4.92". Thickness of Web, 0.79".
Maximum fiber strain = 12000 lbs. per square inch.

Distance between supports, in feet	Safe load, uniformly distributed, (including weight of beam,) in tons of 2000 lbs.	Deflection under this load, in inches.	Weight of beam, in tons of 2000 lbs.	Proper distance, in feet, center to center of beams, for Safe Loads of						
				100 lbs. per sq. ft.	125 lbs. per sq. ft.	150 lbs. per sq. ft.	175 lbs. per sq. ft.	200 lbs. per sq. ft.	250 lbs. per sq. ft.	300 lbs. per sq. ft.
10	15.32	0.13	0.23	30.6	24.5	20.4	17.5	15.3	12.3	10.2
11	13.93	0.16	0.25	25.3	20.3	16.9	14.5	12.7	10.1	8.4
12	12.77	0.19	0.27	21.3	17.0	14.2	12.2	10.6	8.5	7.1
13	11.78	0.22	0.29	18.1	14.5	12.1	10.4	9.1	7.2	6.0
14	10.94	0.26	0.32	15.6	12.5	10.4	8.9	7.8	6.3	5.2
15	10.21	0.30	0.34	13.6	10.9	9.1	7.8	6.8	5.4	4.5
16	9.58	0.34	0.36	12.0	9.6	8.0	6.8	6.0	4.8	4.0
17	9.01	0.38	0.38	10.6	8.5	7.1	6.1	5.3	4.2	3.5
18	8.51	0.43	0.41	9.5	7.6	6.3	5.4	4.7	3.8	3.1
19	8.06	0.48	0.43	8.5	6.8	5.7	4.8	4.2	3.4	2.8
20	7.66	0.53	0.45	7.7	6.1	5.1	4.4	3.8	3.1	2.5
21	7.30	0.58	0.47	7.0	5.6	4.6	4.0	3.5	2.8	2.3
22	6.96	0.64	0.50	6.3	5.1	4.2	3.6	3.2	2.5	2.1
23	6.66	0.70	0.52	5.8	4.6	3.9	3.3	2.9	2.3	1.9
24	6.38	0.76	0.54	5.3	4.2	3.6	3.0	2.7	2.1	1.8
25	6.13	0.82	0.56	4.9	3.9	3.3	2.8	2.5	1.9	1.6
26	5.89	0.89	0.59	4.5	3.6	3.0	2.6	2.3	1.8	1.5
27	5.67	0.96	0.61	4.2	3.4	2.8	2.4	2.1	1.7	1.4
28	5.47	1.03	0.63	3.9	3.1	2.6	2.2	2.0	1.6	1.3
29	5.28	1.11	0.65	3.6	2.9	2.4	2.1	1.8	1.5	1.2
30	5.11	1.19	0.68	3.4	2.7	2.3	1.9	1.7	1.4	1.1
31	4.94	1.27	0.70	3.2	2.6	2.1	1.8	1.6	1.3	1.1
32	4.79	1.35	0.72	3.0	2.4	2.0	1.7	1.5	1.2	1.0
33	4.64	1.44	0.74	2.8	2.2	1.9	1.6	1.4	1.1	.9
34	4.51	1.53	0.77	2.7	2.1	1.8	1.5	1.3	1.1	.9
35	4.38	1.62	0.79	2.5	2.0	1.7	1.4	1.3	1.0	.8
36	4.26	1.71	0.81	2.4	1.9	1.6	1.4	1.2	.9	.8
37	4.14	1.80	0.83	2.2	1.8	1.5	1.3	1.1	.9	.7
38	4.03	1.90	0.86	2.1	1.7	1.4	1.2	1.1	.8	.7
39	3.93	2.01	0.88	2.0	1.6	1.3	1.2	1.0	.8	.7

UNION IRON MILLS'
10-INCH EYEBEAM, No. 5, LIGHT,
30 LBS. PER FOOT.

Depth, 10″. Width of Flanges, 4.32″. Thickness of Web, 0.32″.
Maximum fiber strain = 12000 lbs. per square inch.

Distance between supports, in feet.	Safe load, uniformly distributed, (including weight of beam,) in tons of 2000 lbs.	Deflection under this load, in inches.	Weight of beam, in tons of 2000 lbs.	Proper distance, in feet, center to center of beams, for Safe Loads of						
				100 lbs. per sq. ft.	125 lbs. per sq. ft.	150 lbs. per sq. ft.	175 lbs. per sq. ft.	200 lbs. per sq. ft.	250 lbs. per sq. ft.	300 lbs. per sq. ft.
10	12.00	0.14	0.15	24.0	19.2	16.0	13.7	12.0	9.6	8.0
11	10.91	0.17	0.17	19.8	15.9	13.2	11.3	9.9	7.9	6.6
12	10.00	0.20	0.18	16.7	13.3	11.1	9.5	8.3	6.7	5.6
13	9.23	0.23	0.20	14.2	11.4	9.5	8.1	7.1	5.7	4.7
14	8.57	0.27	0.21	12.2	9.8	8.2	7.0	6.1	4.9	4.1
15	8.00	0.31	0.23	10.7	8.5	7.1	6.1	5.3	4.3	3.6
16	7.50	0.35	0.24	9.4	7.5	6.3	5.4	4.7	3.8	3.1
17	7.06	0.40	0.26	8.3	6.6	5.5	4.7	4.2	3.3	2.8
18	6.67	0.45	0.27	7.4	5.9	4.9	4.2	3.7	3.0	2.5
19	6.32	0.50	0.29	6.7	5.3	4.4	3.8	3.3	2.7	2.2
20	6.00	0.55	0.30	6.0	4.8	4.0	3.4	3.0	2.4	2.0
21	5.71	0.61	0.32	5.4	4.4	3.6	3.1	2.7	2.2	1.8
22	5.45	0.67	0.33	5.0	4.0	3.3	2.8	2.5	2.0	1.7
23	5.22	0.73	0.35	4.5	3.6	3.0	2.6	2.3	1.8	1.5
24	5.00	0.80	0.36	4.2	3.3	2.8	2.4	2.1	1.7	1.4
25	4.80	0.87	0.38	3.8	3.1	2.6	2.2	1.9	1.5	1.3
26	4.62	0.94	0.39	3.6	2.8	2.4	2.0	1.8	1.4	1.2
27	4.44	1.01	0.41	3.3	2.6	2.2	1.9	1.6	1.3	1.1
28	4.29	1.09	0.42	3.1	2.4	2.0	1.7	1.5	1.2	1.0
29	4.14	1.17	0.44	2.9	2.3	1.9	1.6	1.4	1.1	.9
30	4.00	1.25	0.45	2.7	2.1	1.8	1.5	1.3	1.1	.9
31	3.87	1.33	0.47	2.5	2.0	1.7	1.4	1.2	1.0	.8
32	3.75	1.42	0.48	2.3	1.9	1.6	1.3	1.2	.9	.8
33	3.64	1.51	0.50	2.2	1.8	1.5	1.3	1.1	.9	.7
34	3.53	1.60	0.51	2.1	1.7	1.4	1.2	1.0	.8	.7
35	3.43	1.70	0.53	2.0	1.6	1.3	1.1	1.0	.8	.7
36	3.33	1.80	0.54	1.9	1.5	1.2	1.1	.9	.7	.6
37	3.24	1.90	0.56	1.8	1.4	1.2	1.0	.9	.7	.6
38	3.16	2.01	0.57	1.7	1.3	1.1	.9	.8	.7	.6
39	3.08	2.11	0.59	1.6	1.3	1.1	.9	.8	.6	.5

UNION IRON MILLS'
10-INCH EYEBEAM, No. 5, HEAVY,
45 LBS. PER FOOT.

Depth, 10″. Width of Flanges, 4.77″. Thickness of Web, 0.77″.
Maximum fiber strain = 12000 lbs. per square inch.

Distance between supports, in feet.	Safe load, uniformly distributed, (including weight of beam,) in tons of 2000 lbs.	Deflection under this load, in inches.	Weight of beam, in tons of 2000 lbs.	Proper distance, in feet, center to center of beams, for Safe Loads of						
				100 lbs. per sq. ft.	125 lbs. per sq. ft.	150 lbs. per sq. ft.	175 lbs. per sq. ft.	200 lbs. per sq. ft.	250 lbs. per sq. ft.	300 lbs. per sq. ft.
10	15.00	0.14	0.23	30.0	24.0	20.0	17.1	15.0	12.0	10.0
11	13.64	0.17	0.25	24.8	19.8	16.5	14.2	12.4	9.9	8.3
12	12.50	0.20	0.27	20.8	16.7	13.9	11.6	10.4	8.3	6.9
13	11.54	0.23	0.29	17.8	14.2	11.8	10.1	8.9	7.1	5.9
14	10.71	0.27	0.32	15.3	12.2	10.2	8.7	7.7	6.1	5.1
15	10.00	0.31	0.34	13.3	10.7	8.9	7.6	6.7	5.5	4.4
16	9.38	0.35	0.36	11.7	9.4	7.8	6.7	5.9	4.7	3.9
17	8.82	0.40	0.38	10.4	8.3	6.9	5.9	5.2	4.2	3.5
18	8.33	0.45	0.41	9.3	7.4	6.2	5.3	4.6	3.7	3.1
19	7.89	0.50	0.43	8.3	6.6	5.5	4.7	4.2	3.3	2.8
20	7.50	0.55	0.45	7.5	6.0	5.0	4.3	3.8	3.0	2.5
21	7.14	0.61	0.47	6.8	5.4	4.5	3.9	3.4	2.7	2.3
22	6.82	0.67	0.50	6.2	5.0	4.1	3.5	3.1	2.5	2.1
23	6.52	0.73	0.52	5.7	4.5	3.8	3.2	2.8	2.3	1.9
24	6.25	0.80	0.54	5.2	4.1	3.5	2.9	2.6	2.1	1.7
25	6.00	0.87	0.56	4.8	3.8	3.2	2.7	2.4	1.9	1.6
26	5.77	0.94	0.59	4.4	3.6	3.0	2.5	2.2	1.8	1.5
27	5.56	1.01	0.61	4.1	3.3	2.8	2.4	2.1	1.6	1.4
28	5.36	1.09	0.63	3.8	3.1	2.6	2.2	1.9	1.5	1.3
29	5.17	1.17	0.65	3.6	2.9	2.4	2.0	1.8	1.4	1.2
30	5.00	1.25	0.68	3.3	2.7	2.2	1.9	1.7	1.3	1.1
31	4.84	1.33	0.70	3.1	2.5	2.1	1.8	1.6	1.2	1.0
32	4.69	1.42	0.72	2.9	2.3	1.9	1.7	1.5	1.2	1.0
33	4.55	1.51	0.74	2.8	2.2	1.8	1.6	1.4	1.1	.9
34	4.41	1.60	0.77	2.6	2.1	1.7	1.5	1.3	1.0	.9
35	4.29	1.70	0.79	2.4	2.0	1.6	1.4	1.2	1.0	.8
36	4.17	1.80	0.81	2.3	1.9	1.5	1.3	1.2	.9	.8
37	4.05	1.90	0.83	2.2	1.8	1.5	1.3	1.1	.9	.7
38	3.95	2.01	0.86	2.1	1.7	1.4	1.2	1.0	.8	.7
39	3.85	2.11	0.88	2.0	1.6	1.3	1.1	1.0	.8	.7

UNION IRON MILLS'
9-INCH EYEBEAM, No. 6, LIGHT,
23½ LBS. PER FOOT.

Depth, 9″. Width of Flanges, 4.01″. Thickness of Web, 0.26″.
Maximum fiber strain = 12000 lbs. per square inch.

Distance between supports, in feet.	Safe load, uniformly distributed, (including weight of beam,) in tons of 2000 lbs.	Deflection under this load, in inches.	Weight of beam, in tons of 2000 lbs.	Proper distance, in feet, center to center of beams, for Safe Loads of						
				100 lbs. per sq. ft.	125 lbs. per sq. ft.	150 lbs. per sq. ft.	175 lbs. per sq. ft.	200 lbs. per sq. ft.	250 lbs. per sq. ft.	300 lbs. per sq. ft.
10	8.68	0.15	0.12	17.4	13.9	11.6	9.9	8.7	6.9	5.8
11	7.89	0.19	0.13	14.4	11.5	9.6	8.2	7.2	5.7	4.8
12	7.23	0.22	0.14	12.1	9.6	8.0	6.9	6.0	4.8	4.0
13	6.68	0.26	0.15	10.3	8.2	6.9	5.9	5.1	4.1	3.4
14	6.20	0.30	0.16	8.9	7.1	5.9	5.1	4.4	3.5	2.9
15	5.79	0.35	0.18	7.7	6.2	5.1	4.4	3.9	3.1	2.6
16	5.43	0.40	0.19	6.8	5.4	4.5	3.9	3.4	2.7	2.3
17	5.11	0.45	0.20	6.0	4.8	4.0	3.4	3.0	2.4	2.0
18	4.82	0.50	0.21	5.4	4.3	3.6	3.0	2.7	2.1	1.8
19	4.57	0.56	0.22	4.8	3.8	3.2	2.7	2.4	1.9	1.6
20	4.34	0.62	0.24	4.3	3.5	2.9	2.5	2.2	1.7	1.4
21	4.13	0.68	0.25	3.9	3.2	2.6	2.2	2.0	1.6	1.3
22	3.95	0.75	0.26	3.6	2.9	2.4	2.0	1.8	1.4	1.2
23	3.77	0.82	0.27	3.3	2.6	2.2	1.9	1.6	1.3	1.1
24	3.62	0.89	0.28	3.0	2.4	2.0	1.7	1.5	1.2	1.0
25	3.47	0.96	0.29	2.8	2.2	1.9	1.6	1.4	1.1	.9
26	3.34	1.04	0.31	2.6	2.0	1.7	1.5	1.3	1.0	.9
27	3.21	1.12	0.32	2.4	1.9	1.6	1.4	1.2	1.0	.8
28	3.10	1.20	0.33	2.2	1.8	1.5	1.3	1.1	.9	.7
29	2.99	1.29	0.34	2.1	1.6	1.4	1.2	1.0	.8	.7
30	2.89	1.39	0.35	1.9	1.5	1.3	1.1	1.0	.8	.6
31	2.80	1.48	0.36	1.8	1.4	1.2	1.0	.9	.7	.6
32	2.71	1.58	0.38	1.7	1.4	1.1	1.0	.9	.7	.6
33	2.63	1.68	0.39	1.6	1.3	1.1	.9	.8	.6	.5
34	2.55	1.78	0.40	1.5	1.2	1.0	.9	.8	.6	.5
35	2.48	1.89	0.41	1.4	1.1	.9	.8	.7	.6	.5
36	2.41	2.00	0.42	1.3	1.1	.9	.8	.7	.5	.4
37	2.35	2.11	0.43	1.3	1.0	.8	.7	.6	.5	.4
38	2.28	2.22	0.45	1.2	1.0	.8	.7	.6	.5	.4
39	2.23	2.34	0.46	1.2	.9	.8	.7	.6	.5	.4

UNION IRON MILLS'
9-INCH EYEBEAM, No. 6, HEAVY,
33 LBS. PER FOOT.

Depth, 9″. Width of Flanges, 4.33″. Thickness of Web, 0.58″.
Maximum fiber strain = 12000 lbs. per square inch.

Distance between supports, in feet.	Safe load, uniformly distributed, (including weight of beam,) in tons of 2000 lbs.	Deflection under this load, in inches.	Weight of beam, in tons of 2000 lbs.	Proper distance, in feet, center to center of beams, for Safe Loads of							
				100 lbs. per sq. ft.	125 lbs. per sq. ft.	150 lbs. per sq. ft.	175 lbs. per sq. ft.	200 lbs. per sq. ft.	250 lbs. per sq. ft.	300 lbs. per sq. ft.
10	10.40	0.15	0.17	20.8	16.6	13.9	11.9	10.4	8.3	6.9
11	9.45	0.19	0.18	17.2	13.8	11.5	9.8	8.6	6.9	5.7
12	8.67	0.22	0.20	14.5	11.6	9.6	8.3	7.2	5.8	4.8
13	8.00	0.26	0.22	12.3	9.8	8.2	7.0	6.2	4.9	4.1
14	7.43	0.30	0.23	10.6	8.5	7.1	6.1	5.3	4.2	3.5
15	6.93	0.35	0.25	9.2	7.4	6.2	5.3	4.6	3.7	3.1
16	6.50	0.40	0.26	8.1	6.5	5.4	4.6	4.1	3.3	2.7
17	6.12	0.45	0.28	7.2	5.8	4.8	4.1	3.6	2.9	2.4
18	5.78	0.50	0.30	6.4	5.1	4.3	3.7	3.2	2.6	2.1
19	5.47	0.56	0.31	5.8	4.6	3.8	3.3	2.9	2.3	1.9
20	5.20	0.62	0.33	5.2	4.2	3.5	3.0	2.6	2.1	1.7
21	4.95	0.68	0.35	4.7	3.8	3.1	2.7	2.4	1.9	1.6
22	4.73	0.75	0.36	4.3	3.4	2.9	2.5	2.2	1.7	1.4
23	4.52	0.82	0.38	3.9	3.1	2.6	2.3	2.0	1.6	1.3
24	4.33	0.89	0.40	3.6	2.9	2.4	2.1	1.8	1.4	1.2
25	4.16	0.96	0.41	3.3	2.7	2.2	1.9	1.7	1.3	1.1
26	4.00	1.04	0.43	3.1	2.5	2.1	1.8	1.5	1.2	1.0
27	3.85	1.12	0.45	2.9	2.3	1.9	1.6	1.4	1.1	.9
28	3.71	1.20	0.46	2.7	2.1	1.8	1.5	1.3	1.1	.9
29	3.59	1.29	0.48	2.5	2.0	1.6	1.4	1.2	1.0	.8
30	3.47	1.39	0.50	2.3	1.8	1.5	1.3	1.2	.9	.8
31	3.35	1.48	0.51	2.2	1.7	1.4	1.2	1.1	.9	.7
32	3.25	1.58	0.53	2.0	1.6	1.4	1.1	1.0	.8	.7
33	3.15	1.68	0.55	1.9	1.5	1.3	1.1	1.0	.8	.6
34	3.06	1.78	0.56	1.8	1.4	1.2	1.0	.9	.7	.6
35	2.97	1.89	0.58	1.7	1.4	1.1	1.0	.9	.7	.6
36	2.89	2.00	0.59	1.6	1.3	1.1	.9	.8	.6	.5
37	2.81	2.11	0.61	1.5	1.2	1.0	.9	.8	.6	.5
38	2.74	2.22	0.63	1.4	1.2	1.0	.8	.7	.6	.5
39	2.67	2.34	0.64	1.4	1.1	.9	.8	.7	.5	.5

UNION IRON MILLS'
8-INCH EYEBEAM, No. 8, LIGHT, 22 LBS. PER FOOT.

Depth, 8″. Width of Flanges, 3.81″. Thickness of Web, 0.31″.
Maximum fiber strain = 12000 lbs. per square inch.

Distance between supports, in feet.	Safe load, uniformly distributed, (including weight of beam,) in tons of 2000 lbs.	Deflection under this load, in inches.	Weight of beam, in tons of 2000 lbs.	Proper distance, in feet, center to center of beams, for Safe Loads of						
				100 lbs. per sq. ft.	125 lbs. per sq. ft.	150 lbs. per sq. ft.	175 lbs. per sq. ft.	200 lbs. per sq. ft.	250 lbs. per sq. ft.	300 lbs. per sq. ft.
5	14.00	0.04	0.06	56.0	44.8	37.3	32.0	28.0	22.4	18.7
6	11.67	0.06	0.07	38.9	31.1	25.9	22.2	19.5	15.6	13.0
7	10.00	0.08	0.08	28.6	22.9	19.0	16.3	14.3	11.4	9.5
8	8.75	0.11	0.09	21.9	17.5	14.6	12.5	10.9	8.8	7.3
9	7.78	0.14	0.10	17.3	13.8	11.5	9.9	8.6	6.9	5.8
10	7.00	0.17	0.11	14.0	11.2	9.3	8.0	7.0	5.6	4.7
11	6.36	0.21	0.12	11.6	9.2	7.7	6.6	5.8	4.6	3.9
12	5.83	0.25	0.13	9.7	7.8	6.5	5.6	4.9	3.9	3.2
13	5.38	0.29	0.14	8.3	6.6	5.5	4.7	4.1	3.3	2.8
14	5.00	0.34	0.15	7.1	5.7	4.8	4.1	3.6	2.9	2.4
15	4.67	0.39	0.17	6.2	5.0	4.2	3.6	3.1	2.5	2.1
16	4.38	0.44	0.18	5.5	4.4	3.7	3.1	2.7	2.2	1.8
17	4.12	0.50	0.19	4.9	3.9	3.2	2.8	2.4	1.9	1.6
18	3.89	0.56	0.20	4.3	3.5	2.9	2.5	2.2	1.7	1.4
19	3.68	0.62	0.21	3.9	3.1	2.6	2.2	1.9	1.5	1.3
20	3.50	0.69	0.22	3.5	2.8	2.3	2.0	1.8	1.4	1.2
21	3.33	0.76	0.23	3.2	2.5	2.1	1.8	1.6	1.3	1.1
22	3.18	0.84	0.24	2.9	2.3	1.9	1.7	1.4	1.2	1.0
23	3.04	0.92	0.25	2.6	2.1	1.8	1.5	1.3	1.1	.9
24	2.92	1.00	0.26	2.4	1.9	1.6	1.4	1.2	1.0	.8
25	2.80	1.08	0.28	2.2	1.8	1.5	1.3	1.1	.9	.7
26	2.69	1.17	0.29	2.1	1.7	1.4	1.2	1.0	.8	.7
27	2.59	1.26	0.30	1.9	1.5	1.3	1.1	1.0	.8	.6
28	2.50	1.36	0.31	1.8	1.4	1.2	1.0	.9	.7	.6
29	2.41	1.46	0.32	1.7	1.3	1.1	.9	.8	.7	.6
30	2.33	1.56	0.33	1.6	1.2	1.0	.9	.8	.6	.5
31	2.26	1.67	0.34	1.5	1.2	1.0	.8	.7	.6	.5
32	2.19	1.78	0.35	1.4	1.1	.9	.8	.7	.5	.5
33	2.12	1.89	0.36	1.3	1.0	.9	.7	.6	.5	.4
34	2.06	2.00	0.37	1.2	1.0	.8	.7	.6	.5	.4

UNION IRON MILLS'
8-INCH EYEBEAM, No. 8, HEAVY, 35 LBS. PER FOOT.

Depth, 8″. Width of Flanges, 4.29″. Thickness of Web, 0.79″.
Maximum fiber strain = 12000 lbs. per square inch.

Distance between supports, in feet.	Safe load, uniformly distributed, (including weight of beam,) in tons of 2000 lbs.	Deflection under this load, in inches.	Weight of beam, in tons of 2000 lbs.	Proper distance, in feet, center to center of beams, for Safe Loads of						
				100 lbs. per sq. ft.	125 lbs. per sq. ft.	150 lbs. per sq. ft.	175 lbs. per sq. ft.	200 lbs. per sq. ft.	250 lbs. per sq. ft.	300 lbs. per sq. ft.
5	18.08	0.04	0.09	72.3	57.9	48.2	41.3	36.2	28.9	24.1
6	15.07	0.06	0.11	50.2	40.2	33.5	28.7	25.1	20.1	16.7
7	12.91	0.08	0.12	36.9	29.5	24.6	21.1	18.4	14.8	12.3
8	11.30	0.11	0.14	28.3	22.6	18.8	16.1	14.1	11.3	9.4
9	10.04	0.14	0.16	22.3	17.8	14.9	12.7	11.2	8.9	7.4
10	9.04	0.17	0.18	18.1	14.5	12.1	10.3	9.0	7.2	6.0
11	8.22	0.21	0.19	14.9	12.0	10.0	8.5	7.5	6.0	5.0
12	7.53	0.25	0.21	12.6	10.0	8.4	7.2	6.3	5.0	4.2
13	6.95	0.29	0.23	10.7	8.6	7.1	6.1	5.3	4.3	3.6
14	6.46	0.34	0.25	9.2	7.4	6.2	5.3	4.6	3.7	3.1
15	6.03	0.39	0.26	8.0	6.4	5.4	4.6	4.0	3.2	2.7
16	5.65	0.44	0.28	7.1	5.6	4.7	4.0	3.5	2.8	2.4
17	5.32	0.50	0.30	6.3	5.0	4.2	3.6	3.1	2.5	2.1
18	5.02	0.56	0.32	5.6	4.5	3.7	3.2	2.8	2.2	1.9
19	4.76	0.62	0.33	5.0	4.0	3.3	2.9	2.5	2.0	1.7
20	4.52	0.69	0.35	4.5	3.6	3.0	2.6	2.3	1.8	1.5
21	4.30	0.76	0.37	4.1	3.3	2.7	2.3	2.0	1.6	1.4
22	4.11	0.84	0.39	3.7	3.0	2.5	2.1	1.9	1.5	1.2
23	3.93	0.92	0.40	3.4	2.7	2.3	2.0	1.7	1.4	1.1
24	3.77	1.00	0.42	3.1	2.5	2.1	1.8	1.6	1.3	1.0
25	3.62	1.08	0.44	2.9	2.3	1.9	1.7	1.4	1.2	1.0
26	3.48	1.17	0.46	2.7	2.1	1.8	1.5	1.3	1.1	.9
27	3.35	1.26	0.47	2.5	2.0	1.6	1.4	1.2	1.0	.8
28	3.23	1.36	0.49	2.3	1.8	1.5	1.3	1.2	.9	.8
29	3.12	1.46	0.51	2.2	1.7	1.4	1.2	1.1	.9	.7
30	3.01	1.56	0.53	2.0	1.6	1.3	1.1	1.0	.8	.7
31	2.92	1.67	0.54	1.9	1.5	1.2	1.1	.9	.8	.6
32	2.83	1.78	0.56	1.8	1.4	1.2	1.0	.9	.7	.6
33	2.74	1.89	0.58	1.7	1.3	1.1	.9	.8	.7	.6
34	2.66	2.00	0.60	1.6	1.2	1.0	.9	.8	.6	.5

UNION IRON MILLS'

7-INCH EYEBEAM, No. 9, LIGHT, 18 LBS. PER FOOT.

Depth, 7″. Width of Flanges, 3.61″. Thickness of Web, 0.23″.
Maximum fiber strain = 12000 lbs. per square inch.

Distance between supports, in feet.	Safe load, uniformly distributed, (including weight of beam,) in tons of 2000 lbs.	Deflection under this load, in inches.	Weight of beam, in tons of 2000 lbs.	Proper distance, in feet, center to center of beams, for Safe Loads of						
				100 lbs. per sq. ft.	125 lbs. per sq. ft.	150 lbs. per sq. ft.	175 lbs. per sq. ft.	200 lbs. per sq. ft.	250 lbs. per sq. ft.	300 lbs. per sq. ft.
5	10.48	0.05	0.05	41.9	33.5	27.9	24.0	21.0	16.8	14.0
6	8.73	0.07	0.05	29.1	23.3	19.4	16.6	14.6	11.6	9.7
7	7.49	0.10	0.06	21.4	17.1	14.3	12.2	10.7	8.6	7.1
8	6.55	0.13	0.07	16.4	13.1	10.9	9.4	8.2	6.6	5.5
9	5.82	0.16	0.08	12.9	10.3	8.6	7.4	6.5	5.2	4.3
10	5.24	0.20	0.09	10.5	8.4	7.0	6.0	5.2	4.2	3.5
11	4.76	0.24	0.10	8.7	6.9	5.8	4.9	4.3	3.5	2.9
12	4.37	0.28	0.11	7.3	5.8	4.9	4.2	3.6	2.9	2.4
13	4.03	0.33	0.12	6.2	5.0	4.1	3.5	3.1	2.5	2.1
14	3.74	0.39	0.13	5.3	4.3	3.6	3.1	2.7	2.1	1.8
15	3.49	0.45	0.14	4.7	3.7	3.1	2.7	2.3	1.9	1.6
16	3.28	0.51	0.14	4.1	3.3	2.7	2.3	2.1	1.6	1.4
17	3.08	0.57	0.15	3.6	2.9	2.4	2.1	1.8	1.4	1.2
18	2.91	0.64	0.16	3.2	2.6	2.2	1.8	1.6	1.3	1.1
19	2.76	0.71	0.17	2.9	2.3	1.9	1.7	1.5	1.1	1.0
20	2.62	0.79	0.18	2.6	2.1	1.7	1.5	1.3	1.0	.9
21	2.50	0.87	0.19	2.4	1.9	1.6	1.4	1.2	1.0	.8
22	2.38	0.96	0.20	2.2	1.7	1.4	1.2	1.1	.9	.7
23	2.28	1.05	0.21	2.0	1.6	1.3	1.1	1.0	.8	.7
24	2.18	1.14	0.22	1.8	1.4	1.2	1.0	.9	.7	.6
25	2.10	1.24	0.23	1.7	1.3	1.1	1.0	.8	.7	.6
26	2.02	1.34	0.23	1.6	1.2	1.0	.9	.8	.6	.5
27	1.94	1.44	0.24	1.4	1.2	1.0	.8	.7	.6	.5
28	1.87	1.55	0.25	1.3	1.1	.9	.8	.7	.5	.4
29	1.81	1.66	0.26	1.2	1.0	.8	.7	.6	.5	.4

UNION IRON MILLS'

7-INCH EYEBEAM, No. 8, HEAVY, 25 LBS. PER FOOT.

Depth, 7″. Width of Flanges, 3.91″. Thickness of Web, 0.53″.
Maximum fiber strain = 12000 lbs. per square inch.

Distance between supports, in feet.	Safe load, uniformly distributed, (including weight of beam,) in tons of 2000 lbs.	Deflection under this load, in inches.	Weight of beam, in tons of 2000 lbs.	Proper distance, in feet, center to center of beams, for Safe Loads of						
				100 lbs. per sq. ft.	125 lbs. per sq. ft.	150 lbs. per sq. ft.	175 lbs. per sq. ft.	200 lbs. per sq. ft.	250 lbs. per sq. ft.	300 lbs. per sq. ft.
5	12.40	0.05	0.06	49.6	39.7	33.1	28.3	24.8	19.8	16.5
6	10.33	0.07	0.08	34.4	27.5	23.0	19.7	17.2	13.8	11.5
7	8.86	0.10	0.09	25.3	20.2	16.9	14.5	12.7	10.1	8.4
8	7.75	0.13	0.10	19.4	15.5	12.9	11.1	9.7	7.8	6.5
9	6.89	0.16	0.11	15.3	12.2	10.2	8.7	7.7	6.1	5.1
10	6.20	0.20	0.13	12.4	9.9	8.3	7.1	6.2	5.0	4.1
11	5.64	0.24	0.14	10.3	8.2	6.8	5.9	5.1	4.1	3.4
12	5.17	0.28	0.15	8.6	6.9	5.7	4.9	4.3	3.4	2.9
13	4.77	0.33	0.16	7.3	5.9	4.9	4.2	3.7	2.9	2.4
14	4.43	0.39	0.18	6.3	5.1	4.2	3.6	3.2	2.5	2.1
15	4.13	0.45	0.19	5.5	4.4	3.7	3.1	2.8	2.2	1.8
16	3.88	0.51	0.20	4.9	3.9	3.2	2.8	2.4	1.9	1.6
17	3.65	0.57	0.21	4.3	3.4	2.9	2.5	2.1	1.7	1.4
18	3.44	0.64	0.23	3.8	3.1	2.5	2.2	1.9	1.5	1.3
19	3.26	0.71	0.24	3.4	2.7	2.3	2.0	1.7	1.4	1.1
20	3.10	0.79	0.25	3.1	2.5	2.1	1.8	1.5	1.2	1.0
21	2.95	0.87	0.26	2.8	2.2	1.9	1.6	1.4	1.1	.9
22	2.82	0.96	0.28	2.6	2.0	1.7	1.5	1.3	1.0	.9
23	2.70	1.05	0.29	2.4	1.9	1.6	1.3	1.2	.9	.8
24	2.58	1.14	0.30	2.2	1.7	1.4	1.2	1.1	.9	.7
25	2.48	1.24	0.31	2.0	1.6	1.3	1.1	1.0	.8	.7
26	2.38	1.34	0.33	1.8	1.5	1.2	1.0	.9	.7	.6
27	2.30	1.44	0.34	1.7	1.4	1.1	1.0	.9	.7	.6
28	2.21	1.55	0.35	1.6	1.3	1.1	.9	.8	.6	.5
29	2.14	1.66	0.36	1.5	1.2	1.0	.8	.7	.6	.5

UNION IRON MILLS'

6-INCH EYEBEAM, No. 10, LIGHT,
13½ LBS. PER FOOT.

Depth, 6″. Width of Flanges, 3.24″. Thickness of Web, 0.24″.
Maximum fiber strain = 12000 lbs. per square inch.

Distance between supports, in feet.	Safe load, uniformly distributed, (includ-ing weight of beam,) in tons of 2000 lbs.	Deflection under this load, in inches.	Weight of beam, in tons of 2000 lbs.	Proper distance, in feet, center to center of beams, for Safe Loads of						
				100 lbs. per sq. ft.	125 lbs. per sq. ft.	150 lbs. per sq. ft.	175 lbs. per sq. ft.	200 lbs. per sq. ft.	250 lbs. per sq. ft.	300 lbs. per sq. ft.
5	6.53	0.06	0.03	26.1	20.9	17.4	14.9	13.1	10.4	8.7
6	5.44	0.08	0.04	18.1	14.5	12.1	10.4	9.1	7.3	6.0
7	4.66	0.11	0.05	13.3	10.6	8.9	7.6	6.7	5.3	4.4
8	4.08	0.15	0.05	10.2	8.2	6.8	5.8	5.1	4.1	3.4
9	3.63	0.19	0.06	8.1	6.5	5.4	4.6	4.0	3.2	2.7
10	3.26	0.23	0.07	6.5	5.2	4.4	3.7	3.3	2.6	2.2
11	2.97	0.28	0.07	5.4	4.3	3.6	3.1	2.7	2.2	1.8
12	2.72	0.33	0.08	4.5	3.6	3.0	2.6	2.3	1.8	1.5
13	2.51	0.39	0.09	3.9	3.1	2.6	2.2	1.9	1.5	1.3
14	2.33	0.45	0.09	3.3	2.7	2.2	1.9	1.7	1.3	1.1
15	2.18	0.52	0.10	2.9	2.3	1.9	1.7	1.5	1.2	1.0
16	2.04	0.59	0.11	2.6	2.0	1.7	1.5	1.3	1.0	.9
17	1.92	0.67	0.11	2.3	1.8	1.5	1.3	1.1	.9	.8
18	1.81	0.75	0.12	2.0	1.6	1.3	1.1	1.0	.8	.7
19	1.72	0.83	0.13	1.8	1.4	1.2	1.0	.9	.7	.6
20	1.63	0.92	0.14	1.6	1.3	1.1	.9	.8	.7	.5
21	1.55	1.01	0.14	1.5	1.2	1.0	.8	.7	.6	.5
22	1.48	1.11	0.15	1.3	1.1	.9	.8	.7	.5	.5
23	1.42	1.22	0.16	1.2	1.0	.8	.7	.6	.5	.4
24	1.36	1.33	0.16	1.1	.9	.7	.6	.6	.5	.4
25	1.31	1.45	0.17	1.0	.8	.7	.6	.5	.4	.4
26	1.26	1.56	0.18	1.0	.8	.6	.5	.5	.4	.3
27	1.21	1.68	0.18	.9	.7	.6	.5	.4	.4	.3
28	1.17	1.81	0.19	.8	.7	.6	.5	.4	.3	.3
29	1.13	1.95	0.20	.8	.6	.5	.4	.4	.3	.3

UNION IRON MILLS'

6-INCH EYEBEAM, No. 10, HEAVY, 18 LBS. PER FOOT.

Depth, 6″. Width of Flanges, 3.46″. Thickness of Web, 0.46″.
Maximum fiber strain = 12000 lbs. per square inch.

Distance between supports, in feet.	Safe load, uniformly distributed, (including weight of beam,) in tons of 2000 lbs.	Deflection under this load, in inches.	Weight of beam, in tons of 2000 lbs.	Proper distance, in feet, center to center of beams, for Safe Loads of						
				100 lbs. per sq. ft.	125 lbs. per sq. ft.	150 lbs. per sq. ft.	175 lbs. per sq. ft.	200 lbs. per sq. ft.	250 lbs. per sq. ft.	300 lbs. per sq. ft.
5	7.58	0.06	0.05	30.3	24.3	20.2	17.3	15.2	12.1	10.1
6	6.32	0.08	0.05	21.1	16.9	14.0	12.0	10.5	8.4	7.0
7	5.42	0.11	0.06	15.5	12.4	10.3	8.9	7.7	6.2	5.2
8	4.74	0.15	0.07	11.9	9.5	7.9	6.8	5.9	4.7	4.0
9	4.21	0.19	0.08	9.4	7.5	6.2	5.3	4.7	3.7	3.1
10	3.79	0.23	0.09	7.6	6.1	5.1	4.3	3.8	3.0	2.5
11	3.45	0.28	0.10	6.3	5.0	4.2	3.6	3.1	2.5	2.1
12	3.16	0.33	0.11	5.3	4.2	3.5	3.0	2.6	2.1	1.8
13	2.92	0.39	0.12	4.5	3.6	3.0	2.6	2.2	1.8	1.5
14	2.71	0.45	0.13	3.9	3.1	2.6	2.2	1.9	1.5	1.3
15	2.53	0.52	0.14	3.4	2.7	2.2	1.9	1.7	1.3	1.1
16	2.37	0.59	0.14	3.0	2.4	2.0	1.7	1.5	1.2	1.0
17	2.23	0.67	0.15	2.6	2.1	1.7	1.5	1.3	1.0	.9
18	2.11	0.75	0.16	2.3	1.9	1.6	1.3	1.2	.9	.8
19	2.00	0.83	0.17	2.1	1.7	1.4	1.2	1.1	.8	.7
20	1.90	0.92	0.18	1.9	1.5	1.3	1.1	1.0	.8	.6
21	1.81	1.01	0.19	1.7	1.4	1.1	1.0	.9	.7	.6
22	1.72	1.11	0.20	1.6	1.2	1.0	.9	.8	.6	.5
23	1.65	1.22	0.21	1.4	1.1	1.0	.8	.7	.6	.5
24	1.58	1.33	0.22	1.3	1.1	.9	.8	.7	.5	.4
25	1.52	1.45	0.23	1.2	1.0	.8	.7	.6	.5	.4
26	1.46	1.56	0.23	1.1	.9	.7	.6	.6	.4	.4
27	1.40	1.68	0.24	1.0	.8	.7	.6	.5	.4	.3
28	1.35	1.81	0.25	1.0	.8	.6	.5	.5	.4	.3
29	1.31	1.95	0.26	.9	.7	.6	.5	.5	.4	.3

UNION IRON MILLS'

5-INCH EYEBEAM, No. 11, LIGHT, 10 LBS. PER FOOT.

Depth, 5″. Width of Flanges, 2.73″. Thickness of Web, 0.225″.
Maximum fiber strain = 12000 lbs. per square inch.

Distance between supports, in feet.	Safe load, uniformly distributed, (including weight of beam,) in tons of 2000 lbs.	Deflection under this load, in inches.	Weight of beam, in tons of 2000 lbs.	Proper distance, in feet, center to center of beams, for Safe Loads of						
				100 lbs. per sq. ft.	125 lbs. per sq. ft.	150 lbs. per sq. ft.	175 lbs. per sq. ft.	200 lbs. per sq. ft.	250 lbs. per sq. ft.	300 lbs. per sq. ft.
5	3.95	0.07	0.03	15.8	12.6	10.5	9.0	7.9	6.3	5.3
6	3.29	0.10	0.03	11.0	8.8	7.3	6.3	5.5	4.4	3.7
7	2.82	0.14	0.04	8.1	6.4	5.4	4.6	4.0	3.2	2.7
8	2.47	0.18	0.04	6.2	4.9	4.1	3.5	3.1	2.5	2.1
9	2.20	0.23	0.05	4.9	3.9	3.3	2.8	2.4	2.0	1.6
10	1.98	0.28	0.05	4.0	3.2	2.6	2.3	2.0	1.6	1.3
11	1.80	0.34	0.06	3.3	2.6	2.2	1.9	1.7	1.3	1.1
12	1.65	0.40	0.06	2.8	2.2	1.8	1.6	1.4	1.1	.9
13	1.52	0.47	0.07	2.3	1.9	1.6	1.3	1.2	.9	.8
14	1.41	0.55	0.07	2.0	1.6	1.3	1.1	1.0	.8	.7
15	1.32	0.63	0.08	1.8	1.4	1.2	1.0	.9	.7	.6
16	1.24	0.71	0.08	1.6	1.2	1.0	.9	.8	.6	.5
17	1.16	0.80	0.09	1.4	1.1	.9	.8	.7	.5	.5
18	1.10	0.90	0.09	1.2	1.0	.8	.7	.6	.5	.4
19	1.04	1.00	0.10	1.1	.9	.7	.6	.5	.4	.4
20	.99	1.11	0.10	1.0	.8	.7	.6	.5	.4	.3
21	.94	1.22	0.11	.9	.7	.6	.5	.4	.4	.3
22	.90	1.34	0.11	.8	.7	.5	.5	.4	.3	.3
23	.86	1.47	0.12	.7	.6	.5	.4	.4	.3	.2
24	.82	1.60	0.12	.7	.5	.4	.4	.3	.3	.2

UNION IRON MILLS'

5-INCH EYEBEAM, No. 11, HEAVY,
13 LBS. PER FOOT.

Depth, 5″. Width of Flanges, 2.91″. Thickness of Web, 0.405″.
Maximum fiber strain = 12000 lbs. per square inch.

Distance between supports, in feet	Safe load, uniformly distributed, (including weight of beam,) in tons of 2000 lbs.	Deflection under this load, in inches.	Weight of beam, in tons of 2000 lbs.	Proper distance, in feet, center to center of beams, for Safe Loads of						
				100 lbs. per sq. ft.	125 lbs. per sq. ft.	150 lbs. per sq. ft.	175 lbs. per sq. ft.	200 lbs. per sq. ft.	250 lbs. per sq. ft.	300 lbs. per sq. ft.
5	4.55	0.07	0.03	18.2	14.6	12.1	10.4	9.1	7.3	6.1
6	3.79	0.10	0.04	12.6	10.1	8.4	7.2	6.3	5.1	4.2
7	3.25	0.14	0.05	9.3	7.4	6.2	5.3	4.6	3.7	3.1
8	2.85	0.18	0.05	7.1	5.7	4.8	4.1	3.6	2.9	2.4
9	2.53	0.23	0.06	5.6	4.5	3.7	3.2	2.8	2.2	1.9
10	2.28	0.28	0.07	4.6	3.6	3.0	2.6	2.3	1.8	1.5
11	2.07	0.34	0.07	3.8	3.0	2.5	2.1	1.9	1.5	1.3
12	1.90	0.40	0.08	3.2	2.5	2.1	1.8	1.6	1.3	1.1
13	1.75	0.47	0.08	2.7	2.2	1.8	1.5	1.3	1.1	.9
14	1.63	0.55	0.09	2.3	1.9	1.6	1.3	1.2	.9	.8
15	1.52	0.63	0.10	2.0	1.6	1.4	1.2	1.0	.8	.7
16	1.42	0.71	0.10	1.8	1.4	1.2	1.0	.9	.7	.6
17	1.34	0.80	0.11	1.6	1.3	1.0	.9	.8	.6	.5
18	1.26	0.90	0.12	1.4	1.1	.9	.8	.7	.6	.5
19	1.20	1.00	0.12	1.3	1.0	.8	.7	.6	.5	.4
20	1.14	1.11	0.13	1.1	.9	.8	.7	.6	.5	.4
21	1.08	1.22	0.14	1.0	.8	.7	.6	.5	.4	.3
22	1.03	1.34	0.14	.9	.8	.6	.5	.5	.4	.3
23	.99	1.47	0.15	.9	.7	.6	.5	.4	.3	.3
24	.95	1.60	0.16	.8	.6	.5	.5	.4	.3	.3

UNION IRON MILLS'

4-INCH EYEBEAM, No. 12, LIGHT, 8 LBS. PER FOOT.

Depth, 4″. Width of Flanges, 2.48″. Thickness of Web, 0.23″.
Maximum fiber strain = 12000 lbs. per square inch.

Distance between supports, in feet.	Safe load, uniformly distributed, (including weight of beam,) in tons of 2000 lbs.	Deflection under this load, in inches.	Weight of beam, in tons of 2000 lbs.	Proper distance, in feet, center to center of beams, for Safe Loads of						
				100 lbs. per sq. ft.	125 lbs. per sq. ft.	150 lbs. per sq. ft.	175 lbs. per sq. ft.	200 lbs. per sq. ft.	250 lbs. per sq. ft.	300 lbs. per sq. ft.
5	2.48	0.09	0.02	9.9	7.9	6.6	5.7	5.0	4.0	3.3
6	2.07	0.13	0.02	6.9	5.5	4.6	3.9	3.5	2.8	2.3
7	1.77	0.17	0.03	5.1	4.0	3.4	2.9	2.5	2.0	1.7
8	1.55	0.22	0.03	3.9	3.1	2.6	2.2	1.9	1.6	1.3
9	1.38	0.28	0.04	3.1	2.5	2.0	1.8	1.5	1.2	1.0
10	1.24	0.35	0.04	2.5	2.0	1.7	1.4	1.2	1.0	.8
11	1.13	0.42	0.04	2.1	1.6	1.4	1.2	1.0	.8	.7
12	1.03	0.50	0.05	1.7	1.4	1.1	1.0	.9	.7	.6
13	0.95	0.59	0.05	1.5	1.2	1.0	.8	.7	.6	.5
14	0.89	0.68	0.06	1.3	1.0	.8	.7	.6	.5	.4
15	0.83	0.78	0.06	1.1	.9	.7	.6	.6	.4	.4
16	0.78	0.89	0.06	1.0	.8	.6	.6	.5	.4	.3
17	0.73	1.01	0.07	.9	.7	.6	.5	.4	.3	.3
18	0.69	1.13	0.07	.8	.6	.5	.4	.4	.3	.3
19	0.65	1.26	0.08	.7	.5	.5	.4	.3	.3	.2

UNION IRON MILLS'

4-INCH EYEBEAM, No. 12, HEAVY,
10 LBS. PER FOOT.

Depth, 4″. Width of Flanges, 2.63″. Thickness of Web, 0.38″.
Maximum fiber strain = 12000 lbs. per square inch.

Distance between supports, in feet.	Safe load, uniformly distributed, (including weight of beam,) in tons of 2000 lbs.	Deflection under this load, in inches.	Weight of beam, in tons of 2000 lbs.	Proper distance, in feet, center to center of beams, for Safe Loads of						
				100 lbs. per sq. ft.	125 lbs. per sq. ft.	150 lbs. per sq. ft.	175 lbs. per sq. ft.	200 lbs. per sq. ft.	250 lbs. per sq. ft.	300 lbs. per sq. ft.
5	2.80	0.09	0.03	11.2	9.0	7.5	6.4	5.6	4.5	3.7
6	2.33	0.13	0.03	7.8	6.2	5.2	4.4	3.9	3.1	2.6
7	2.00	0.17	0.04	5.7	4.6	3.8	3.3	2.9	2.3	1.9
8	1.75	0.22	0.04	4.4	3.5	2.9	2.5	2.2	1.8	1.5
9	1.56	0.28	0.05	3.5	2.8	2.3	2.0	1.7	1.4	1.2
10	1.40	0.35	0.05	2.8	2.2	1.9	1.6	1.4	1.1	.9
11	1.27	0.42	0.06	2.3	1.8	1.5	1.3	1.2	.9	.8
12	1.17	0.50	0.06	2.0	1.6	1.3	1.1	1.0	.8	.7
13	1.08	0.59	0.07	1.7	1.3	1.1	.9	.8	.7	.6
14	1.00	0.68	0.07	1.4	1.1	1.0	.8	.7	.6	.5
15	0.93	0.78	0.08	1.2	1.0	.8	.7	.6	.5	.4
16	0.88	0.89	0.08	1.1	.9	.7	.6	.6	.4	.4
17	0.82	1.01	0.09	1.0	.8	.6	.6	.5	.4	.3
18	0.78	1.13	0.09	.9	.7	.6	.5	.4	.3	.3
19	0.74	1.26	0.10	.8	.6	.5	.4	.4	.3	.3

UNION IRON MILLS'
3-INCH EYEBEAM, No. 13, LIGHT, 7 LBS. PER FOOT.

Depth, 3″. Width of Flanges, 2.32″. Thickness of Web, 0.19″.
Maximum fiber strain = 12000 lbs. per square inch.

Distance between supports, in feet.	Safe load, uniformly distributed, (including weight of beam,) in tons of 2000 lbs.	Deflection under this load, in inches.	Weight of beam, in tons of 2000 lbs.	Proper distance, in feet, center to center of beams, for Safe Loads of						
				100 lbs. per sq. ft.	125 lbs. per sq. ft.	150 lbs. per sq. ft.	175 lbs. per sq. ft.	200 lbs. per sq. ft.	250 lbs. per sq. ft.	300 lbs. per sq. ft.
5	1.65	0.12	0.02	6.6	5.3	4.4	3.8	3.3	2.6	2.2
6	1.37	0.17	0.02	4.6	3.7	3.0	2.6	2.3	1.8	1.5
7	1.18	0.23	0.02	3.4	2.7	2.2	1.9	1.7	1.3	1.1
8	1.03	0.29	0.03	2.6	2.1	1.7	1.5	1.3	1.0	.9
9	0.92	0.37	0.03	2.0	1.6	1.4	1.2	1.0	.8	.7
10	0.82	0.46	0.04	1.6	1.3	1.1	.9	.8	.7	.5
11	0.75	0.56	0.04	1.4	1.1	.9	.8	.7	.5	.5
12	0.69	0.67	0.04	1.2	.9	.8	.7	.6	.5	.4
13	0.63	0.78	0.05	1.0	.8	.6	.6	.5	.4	.3
14	0.59	0.91	0.05	.8	.7	.6	.5	.4	.3	.3

UNION IRON MILLS'
3-INCH EYEBEAM, No. 13, HEAVY, 9 LBS. PER FOOT.

Depth, 3″. Width of Flanges, 2.52″. Thickness of Web, 0.39″.
Maximum fiber strain = 12000 lbs. per square inch.

Distance between supports, in feet.	Safe load, uniformly distributed, (including weight of beam,) in tons of 2000 lbs.	Deflection under this load, in inches.	Weight of beam, in tons of 2000 lbs.	Proper distance, in feet, center to center of beams, for Safe Loads of						
				100 lbs. per sq. ft.	125 lbs. per sq. ft.	150 lbs. per sq. ft.	175 lbs. per sq. ft.	200 lbs. per sq. ft.	250 lbs. per sq. ft.	300 lbs. per sq. ft.
5	1.89	0.12	0.02	7.6	6.0	5.0	4.3	3.8	3.0	2.5
6	1.57	0.17	0.03	5.2	4.2	3.5	3.0	2.6	2.1	1.7
7	1.35	0.23	0.03	3.9	3.1	2.6	2.2	1.9	1.5	1.3
8	1.18	0.29	0.04	3.0	2.4	2.0	1.7	1.5	1.2	1.0
9	1.05	0.37	0.04	2.3	1.9	1.6	1.3	1.2	.9	.8
10	0.94	0.46	0.05	1.9	1.5	1.3	1.1	.9	.8	.6
11	0.86	0.56	0.05	1.6	1.2	1.0	.9	.8	.6	.5
12	0.79	0.67	0.05	1.3	1.1	.9	.8	.7	.5	.4
13	0.73	0.78	0.06	1.1	.9	.7	.6	.6	.4	.4
14	0.67	0.91	0.06	1.0	.8	.6	.5	.5	.4	.3

EXPLANATION OF TABLES ON THE

PROPERTIES OF UNION IRON MILLS' EYE AND DECK BEAMS, CHANNEL BARS, ANGLE, STAR AND TEE IRONS.

Pages 62 to 69, inclusive.

The tables on I Beams, Deck Beams and Channel Bars are calculated for the minimum and maximum weight to which the various shapes can be rolled. The lithographed plates indicate the manner in which the enlargement of the section takes place, and column 7 in tables gives the increase of thickness of web for each pound increase of weight of beam or channel. The width of flanges is increased the same amount as the thickness of web.

Angle Irons are increased in weight in the manner indicated by Fig. 4 on page 23, the size corresponding with the least thickness, and increasing somewhat with the increase of thickness, but some of the heavier weights of a few of the shapes are rolled in special finishing grooves, whereby the exact size is obtained for a thickness greater than the minimum. In the tables, for the sake of uniformity, it was assumed generally that the size corresponds with the least thickness only, and the increase of weight is obtained in the manner indicated by the above mentioned Fig. 4, page 23.

Beams, Channels and Angle Irons, may be rolled to any weight intermediate between the minimum and maximum weights given. Each shape of Star and T Iron, however, can be rolled to one weight only.

Columns 11 and 13 in the tables for beams and channels give coefficients, by the help of which the safe uniformly distributed load for any beam or channel, and for any span length, can be readily and quickly determined. To do this, it is only necessary to divide the coefficient given by the span or distance between supports, in feet, and multiply by 1000. If the weight of the beam or channel is intermediate between the minimum and maximum weights given, add to the coefficient for the minimum weight, the value given in columns 12 or 14 (for one pound increase of weight) multiplied by the number of pounds the beam or channel is heavier than the minimum.

If a beam or channel is to be selected, (as will usually be the case,) intended to carry a certain load for a length of span already determined on, it will be most convenient to ascertain the coefficient which this load and span will require, and refer to the table for a beam or channel having a coefficient as large as this. The coefficient is obtained by multiplying the load, in pounds uniformly distributed, by the span length in feet, and dividing by 1000.

In case the load is not uniformly distributed, but is concentrated at the middle of the beam or channel, multiply the load by 2, and then consider it as uniformly distributed. The deflection will be $\frac{8}{10}$ths of the deflection by this load.

If the load is neither uniformly distributed nor concentrated at the middle, obtain the bending moment. This, multiplied by 0.008 will give the required coefficient.

If the loads for which the beams or channels are to be proportioned, are quiescent, the coefficients for a fiber strain of 12000 lbs. per square inch should be used; but if moving loads are to be provided for, the coefficients for 10000 lbs. fiber strain should be taken. Inasmuch as the effects of impact may be very considerable, (the strains produced in an unyielding inelastic material by a load suddenly applied, being double those produced by the same load in a quiescent condition,) it will sometimes be advisable to use still smaller fiber strains than 10000. The coefficients for these can readily be determined by proportion. Thus, for a fiber strain of 8000 lbs. per square inch, the coefficient will equal the coefficient for 10000 lbs. fiber strain multiplied by $\frac{8}{10}$ths.

The table on the properties of Union Iron Mills' Angle Irons requires explanation only relative to the angles with unequal legs, to which the latter half of the table applies. It will be observed that two values are given, in the case of each angle, for the distance of center of gravity from outside of flange, the moment of inertia, the moment of resistance and the radius of gyration of the section. The first or larger value invariably refers to a neutral axis parallel to the *smaller* flange, and to the distance between the center of gravity and the outside of this flange, and the second or smaller value to a neutral axis parallel to the *larger* flange, and to the distance between the center of gravity and the outside of this flange. For each position of the neutral axis there will be two moments of resistance, since the distance between the neutral axis and the extreme fibers has a different value on one side of the axis from what it has on the other. The moment of resistance given in table is the smaller of these two values, and the fiber strain calculated from it, will therefore give the larger of the two strains in extreme fibers, (since these strains are equal to the bending moment divided by the moment of resistance of the section). The left hand figures in each column refer to the minimum weight of angle, and the right hand figures to the maximum weight, throughout the table.

The table on the properties of Union Iron Mills' T Irons is modeled after the foregoing, and will therefore scarcely require explanation. The horizontal portion of the T is called the flange and the vertical portion the stem. In the case of the neutral axis parallel to the flange, there will be two moments of resistance, and the least is given; but in the case of the neutral axis coincident with stem, there is only one moment of resistance. In calculating the table, the flange and stem of the T's were considered as rectangles of equal area as the actual section, and the figures given are therefore approximations only, though very close ones.

No approximations have entered into the calculations of any of the other tables, and the figures given may be relied upon as accurate.

The use of the radii of gyration will be explained in connection with the table on the strength of wrought iron columns. The moment of resistance is used to determine the fiber strain in a beam or other shape iron subjected to bending or transverse strains, by simply dividing the same into the bending moment, expressed in inch pounds.

The 15th column in the table on the Properties of Union Iron Mills' Channels, giving the distance of the center of gravity of channels from outside of web, is used to obtain the radius of gyration for columns or struts consisting of two channels latticed, as represented by Fig. 1, page 26, in the case of the neutral axis passing through the center of the section parallel to the webs of the channels. This radius of gyration is equal to the square root of the distance between the center of gravity of the channel and the center of the section.

EXAMPLES OF APPLICATION OF TABLES.

I. What load, uniformly distributed, will a 10″ beam carry, weighing 40 lbs. per foot, and measuring 14 feet between supports, allowing a fiber strain of 12000 lbs. per square inch?

Answer: By table, C, for a 10″ beam, 40 lbs., $= 240 + 10 \times 4 = 280$, therefore $L = \dfrac{1000 \times 280}{14} = 20000$ lbs., including weight of beam.

II. What beam will be required to carry 36000 lbs., uniformly distributed over a span of 16 feet between supports, same fiber strain?

Answer: C required $= \dfrac{IL}{1000} = \dfrac{16 \times 36000}{1000} = 576$, which calls for a 15″ beam, 52 lbs. per foot.

III. A light 4″ × 3″ angle iron, weighing 8.3 lbs. per foot, spanning 4 feet, is loaded with 1000 lbs. at center: what will be the maximum fiber strain if the 4″ flange is in a vertical position?

Answer: By table, moment of resistance $= 1.46$. Bending moment $= 12000$ inch pounds. Therefore maximum fiber strain $= \dfrac{12000}{1.46} = 8220$ lbs., occurring in the fibers furthest from the neutral axis, *i. e.*, at the end of the long flange.

SPECIAL CASES OF LOADING.

I. Beam loaded at a point distant "a" feet from the left hand and "b" feet from the right hand support, by a single load P.

l $=$ length of beam between supports $=$ a $+$ b.

Maximum bending moment, neglecting dead weight of beam, occurs at point of application of the load and $= \dfrac{P\,ab}{l}$

$P =$ load given in tables $\times \dfrac{l^2}{8\,ab}$

Pressure or *reaction* at left hand support $= P\dfrac{b}{l}$, and at right hand support $= P\dfrac{a}{l}$.

II. Beam unsupported at one end and held horizontally at the other, l representing the length of beam from end to support.

If loaded by a uniformly distributed load W:

Maximum bending moment occurs at support and $= \dfrac{W\,l}{2}$

$W =$ load given in tables $\times \frac{1}{4}$, and the deflection $=$ that of the tables $\times 2.4$.

If loaded with a single load P at its extremity:

Maximum bending moment occurs at support and $= Pl$.

$P =$ load given in tables $\times \frac{1}{8}$, and the deflection that of tables $\times 3.2$.

GENERAL FORMULÆ ON THE FLEXURE OF BEAMS OF ANY CROSS-SECTION.

Let $A =$ area of section,
$l =$ length of span,
$W =$ load, uniformly distributed,
$M =$ bending moment,
$d =$ depth of beam, out to out,
$n =$ distance of center of gravity of section, from top or from bottom,
$s =$ strain per square inch in extreme fibers of beam, either top or bottom,
$D =$ maximum deflection,
$I =$ moment of inertia of section,
$R =$ moment of resistance,
$r =$ radius of gyration,
$E =$ modulus of elasticity,
(assumed $= 26000000$ for wrought iron in tables.)

Then $R = \dfrac{I}{n}$; $r = \sqrt{\dfrac{I}{A}}$,

$M = \dfrac{sI}{n} = sR$,

$s = \dfrac{Mn}{I} = \dfrac{M}{R}$,

$W = \dfrac{8\,sI}{ln} = \dfrac{8\,s}{l}R$,

$s = \dfrac{Wln}{8\,I} = \dfrac{Wl}{8\,R}$,

$D = \dfrac{5\,Wl^3}{384\,EI}$ for beam supported at both ends and uniformly loaded,

$D = \dfrac{Pl^3}{48\,EI}$ for beam supported at both ends and loaded by a single load P at middle,

$D = \dfrac{Wl^3}{8\,EI}$ for beam held horizontally at one end only and uniformly loaded,

$D = \dfrac{Pl^3}{3\,EI}$ for beam held horizontally at one end only and loaded with a single load P at the other.

VALUES OF I AND R FOR USUAL SECTIONS.

Rectangle; h = hight, b = base; for neutral axis through center of gravity, parallel to base, $I = \dfrac{bh^3}{12}$, $R = \dfrac{bh^2}{6}$; for neutral axis coincident with base, $I = \dfrac{bh^3}{3}$.

Triangle; h = hight, b = base; for neutral axis through center of gravity (*i. e.*, distant ⅓ h from base), parallel to base, $I = \dfrac{bh^3}{36}$, $R_{min.} = \dfrac{bh^2}{24}$; for neutral axis coincident with base, $I = \dfrac{bh^3}{12}$; for neutral axis through apex, parallel to base, $I = \dfrac{bh^3}{4}$.

Circle; d = diameter, π = 3.1416; for neutral axis through center, $I = \dfrac{\pi d^4}{64} = 0.0491\,d^4$, $R = \dfrac{\pi d^3}{32} = 0.0982\,d^3$.

PROPERTIES OF UNION IRON MILLS' EYEBEAMS.

1	2	3	4	5	6	7	8	9	10	11	12	13	14	15	16
No. of Shape.	Designation.	Weight per foot.	Area of Section.	Thickness of Web.	Width of Flange.	Increase of thickness of web for each lb. increase of weight.	Moment of Inertia, neutral axis perpendicular to web at center.	Moment of Resistance, neutral axis as before.	Radius of Gyration, neutral axis as before.	C For fiber strain of 12000 lbs. per sq. in.	Add for every lb. increase of weight of beam	C For fiber strain of 10000 lbs. per sq. in.	Add for every lb. increase of weight of beam	Moment of Inertia, neutral axis coincident with center line of web.	Radius of Gyration, neutral axis as before.
		lbs.	Sq. in.	Inches.	Inches.	Inches.									
1	15″ Light,	50.	15.0	.47	5.03	.02	530.	70.6	5.94	585.	6.0	471.	5.0	16.3	1.04
1	15″ Heavy,	65.	19.5	.77	5.33		614.	81.9	5.61	655.	6.0	546.	5.0	20.0	1.01
2	15″ Light,	67.	20.1	.67	5.55	.02	677.	90.3	5.80	722.		602.		25.4	1.12
2	15″ Heavy,	80.	24.0	.93	5.81		750.	100.	5.59	800.		667.		29.9	1.12
3	12″ Light,	42.	12.6	.51	4.64	.025	275.	45.9	4.68	367.	4.8	306.	4.0	11.0	0.94
3	12″ Heavy,	60.	18.0	.96	5.09		340.	56.7	4.35	454.		378.		15.5	0.93
4	10½″ Light,	31.5	9.5	.41	4.54	.029	165.	31.4	4.17	251.	4.1	209.	3.4	8.01	0.92
4	10½″ Heavy,	45.	13.5	.79	4.92		201.	38.3	3.86	306.		255.		10.7	0.89
5	10″ Light,	30.	9.0	.32	4.32	.03	150.	30.0	4.09	240.	4.0	200.	3.3	7.94	0.94
5	10″ Heavy,	45.	13.5	.77	4.77		187.	37.5	3.73	300.		250.		11.3	0.91
6	9″ Light,	23.5	7.0	.26	4.01	.033	97.5	21.7	3.73	174.	3.7	145.	2.9	5.48	0.88
6	9″ Heavy,	33.	9.9	.58	4.33		117.	26.0	3.44	208.		173.		7.14	0.85

$L = \dfrac{1000\,C}{M}$ $C = \dfrac{1L}{1000}$

$M = 125\,C.$ $C = .008\,M.$

L = Safe load in lbs. uniformly distrib'd.
C = Coefficient given below.
M = Moment of forces, in foot-lbs.
l = Span in feet.

1	2	3	4	5	6	7	8	9	10	11	12	13	14	15	16
7	9″ Light,Extra,	45.	13.5	.75	4.94	.033	159.	35.3	3.42	282.	3.5	235.	3.0	14.0	1.01
7	9″ Heavy, "	50.	15.0	.91	5.10		169.	37.5	3.34	300.		250.		15.7	1.02
8	8″ Light,	22.	6.6	.31	3.81	.038	69.9	17.5	3.25	140.	3.1	117.	2.6	4.57	0.83
8	8″ Heavy,	35.	10.5	.79	4.29		90.4	22.6	2.94	181.		151.		6.96	0.82
9	7″ Light,	18.	5.4	.23	3.61	.043	45.8	13.1	2.91	105.	2.7	87.3	2.3	3.72	0.83
9	7″ Heavy,	25.	7.5	.53	3.91	.05	54.3	15.5	2.69	124.		103.		4.87	0.81
10	6″ Light,	13.5	4.1	.24	3.24		24.5	8.16	2.46	65.3	2.3	54.4	2.0	2.00	0.70
10	6″ Heavy,	18.	5.4	.46	3.46		28.4	9.48	2.30	75.8		63.2		2.51	0.68
11	5″ Light,	10.	3.0	.225	2.73	.06	12.3	4.94	2.03	39.5	2.0	32.9	1.7	1.08	0.60
11	5″ Heavy,	13.	3.9	.405	2.91	.075	14.2	5.69	1.91	45.5		37.9		1.34	0.59
12	4″ Light,	8.	2.4	.23	2.48		6.19	3.10	1.61	24.8	1.6	20.7	1.3	0.71	0.55
12	4″ Heavy,	10.	3.0	.38	2.63		6.99	3.50	1.53	28.0		23.3		0.87	0.54
13	3″ Light,	7.	2.1	.19	2.32	.1	3.09	2.06	1.21	16.5	1.2	13.7	1.0	0.55	0.55
13	3″ Heavy,	9.	2.7	.39	2.52		3.54	2.36	1.15	18.9		15.7		0.84	0.56

DECK BEAMS.

1	2	3	4	5	6	7	8	9	10	11	12	13	14	15	16
20	9″ Light,	23.5	7.1	.406	3.75	.033	78.6	17.1	3.34	137.	3.6	114.	3.0	2.49	0.59
20	9″ Heavy,	30.	9.0	.625	3.97		81.9	20.0	3.20	160.		133.		3.17	0.59
21	8″ Light,	21.5	6.5	.500	3.75	.038	52.1	11.6	2.84	92.8	3.1	77.3	2.6	2.23	0.59
21	8″ Heavy,	28.	8.4	.750	4.00		63.3	14.1	2.74	112.8		94.0		2.96	0.59
22	7″ Light,	17.	5.1	.375	3.50	.043	34.4	8.6	2.60	68.8	2.9	57.3	2.5	1.81	0.60
22	7″ Heavy,	23.	6.9	.625	3.75		43.0	10.8	2.50	86.4		72.0		2.41	0.59

PROPERTIES OF UNION IRON MILLS' CHANNEL BARS.

$L = \text{Safe load in lbs. uniformly distrib'd.}$
$C = \text{Coefficient given below.}$
$M = \text{Moment of forces, in foot-lbs.}$
$1 = \text{Span in feet.}$

$L = \dfrac{1000 \, C}{1}$ $\quad C = \dfrac{1L}{1000}$

$M = 125 \, C.$ $\quad C = .008 \, M.$

1	2	3	4	5	6	7	8	9	10	11	12	13	14	15
No. of Shape.	Designation.	Weight per foot.	Area of Section.	Thickness of Web.	Width of Flange.	Increase of thickness of web for each lb. increase of weight.	Moment of Inertia, neutral axis perpendicular to web at center.	Moment of Resistance, neutral axis as before.	Radius of Gyration, neutral axis as before.	For fiber strain of 12000 lbs. per sq. in.	Add for every lb. increase of weight of beam	For fiber strain of 10000 lbs. per sq. in.	Add for every lb. increase of weight of beam	Dist. of center of gravity from outside of web.
		Lbs.	Sq. in.	Inches.	Inches.	Inches.								Inches.
25	15″ Light,	40.	12.00	.525	3.53	.0200	359.	47.8	5.47	382.	6.0	319.	5.0	.82
25	15″ Heavy,	60.	18.00	.925	3.93		471.	62.8	5.12	502.		419.		.88
26	12″ One weight	20.	6.00	.318	3.01	.0250	119.	19.9	4.46	159.	4.8	133.	4.0	.69
27	12″ Light,	22.5	6.75	.324	3.01		140.	23.4	4.56	187.		156.		.74
27	12″ Heavy,	30.	9.00	.512	3.20	.0250	168.	27.9	4.31	223.	4.8	186.	4.0	.72
28	12″ Light,	30.	9.00	.457	2.71		176.	29.4	4.42	235.		196.		.72
28	12″ Heavy,	50.	15.00	.957	3.21		248.	41.4	4.07	331.		276.		.83
29	10″ One weight	16.	4.80	.329	2.52	.0300	62.5	12.5	3.61	100.	4.0	83.3	3.4	.55
30	10″ Light,	17.5	5.25	.300	2.43		75.5	15.1	3.79	121.		100.7		.63
30	10″ Heavy,	30.	9.00	.675	2.80	.0300	106.8	21.4	3.44	171.	4.0	142.7	3.3	.66
31	10″ Light,	20.	6.00	.305	2.56		89.4	17.9	3.86	143.		119.3		.70
31	10″ Heavy,	35.	10.50	.755	3.01		126.9	25.4	3.48	203.	4.0	169.3		.65

#											
32	9″ One weight	14.5	4.35	.316	2.50		47.4	10.5	3.30	84.0		70.0	3.6	.58
33	9″ Light,	13.	5.40	.305	2.43	.0333	64.8	14.4	3.46	115.2		96.0	3.0	.68
33	9″ Heavy,	30.	9.00	.705	2.83		89.1	19.8	3.15	158.4		132.0		.73
34	8″ Light,	12.5	3.75	.264	2.01	.0375	34.5	8.61	3.03	68.9	3.2	57.4	2.7	.53
34	8″ Heavy,	15.5	4.65	.376	2.13		39.2	9.81	2.90	78.5		65.4		.53
35	8″ Light,	16.	4.80	.303	2.30	.0375	45.3	11.34	3.07	90.4	3.2	75.6	2.6	.66
35	8″ Heavy,	28.	8.40	.753	2.75		64.5	16.14	2.77	129.1		107.6		.73
36	7″ Light,	10.5	3.15	.247	2.00	.0429	22.4	6.41	2.67	51.3	2.8	42.7	2.3	.52
36	7″ Heavy,	13.5	4.05	.375	2.13		26.1	7.46	2.54	59.7		49.7		.52
37	7″ Light,	14.	4.20	.296	2.30	.0429	30.6	8.73	2.70	69.8	2.8	58.2	2.3	.66
37	7″ Heavy,	20.	6.00	.554	2.55		37.9	10.83	2.51	86.4		72.2		.68
38	6″ Light,	7.5	2.25	.196	1.76	.0500	12.1	4.04	2.32	32.3	2.4	26.9	2.0	.48
38	6″ Heavy,	9.5	2.85	.296	1.86		13.9	4.64	2.21	37.1		30.9		.47
39	6″ Light,	10.	3.00	.227	1.98	.0500	16.6	5.53	2.35	44.2	2.4	36.9	2.0	.60
39	6″ Heavy,	16.	4.80	.527	2.28		22.0	7.33	2.14	58.6		48.9		.62
40	5″ Light,	6.5	1.95	.219	1.66	.0600	7.00	2.80	1.90	22.4	2.0	18.7	1.7	.44
40	5″ Heavy,	8.5	2.55	.339	1.78		8.25	3.30	1.80	26.4		22.0		.44
41	5″ Light,	9.	2.70	.245	1.93	.0600	10.22	4.09	1.94	32.7	2.0	27.3	1.7	.61
41	5″ Heavy,	14.	4.20	.545	2.23		13.35	5.34	1.78	42.7		35.6		.64

PROPERTIES OF UNION IRON MILLS' ANGLE IRONS OF MINIMUM AND MAXIMUM THICKNESSES AND WEIGHTS.

ANGLES WITH EQUAL LEGS.

Size. Inches.	Thickness. Inches.		Weight per Foot. Lbs.		Area. Square Inches.		Dist. of center of gravity from outside of flange. Inches.		Moment of Inertia, neutral axis through center of gravity parallel to flange.		Moment of Resistance, neutral axis as before.		Radius of Gyration, neutral axis as before. Inches.	
	Min.	Max.	Min.	Max.	Min.	Max.	Min.	Max.	Min.	Max.	Min.	Max.	Min.	Max.
6 × 6	1/2	1	19.2	39.2	5.75	11.75	1.68	1.96	19.9	43.1	4.6	9.5	1.9	1.9
4 × 4	3/8	3/4	9.5	19.5	2.86	5.86	1.14	1.35	4.36	9.55	1.5	3.2	1.2	1.3
3 1/2 × 3 1/2	3/8	3/4	8.3	17.0	2.48	5.11	1.01	1.22	2.87	6.38	1.2	2.4	1.1	1.1
3 1/4 × 3 1/4	3/8	3/4	7.7	15.8	2.30	4.73	0.95	1.16	2.27	5.10	0.99	2.1	0.99	1.0
3 × 3	5/16	5/8	5.9	12.2	1.78	3.65	0.86	1.04	1.51	3.35	0.71	1.5	0.92	0.96
2 3/4 × 2 3/4	5/16	1/2	5.4	8.8	1.62	2.65	0.80	0.91	1.15	1.99	0.59	0.98	0.84	0.87
2 1/2 × 2 1/2	5/16	1/2	4.9	8.0	1.46	2.39	0.74	0.85	0.85	1.49	0.48	0.81	0.76	0.79
2 1/4 × 2 1/4	1/4	1/2	3.5	7.3	1.06	2.19	0.66	0.79	0.50	1.13	0.32	0.66	0.69	0.72
2 × 2	1/4	7/16	3.1	5.6	0.94	1.69	0.59	0.70	0.35	0.68	0.25	0.45	0.61	0.63
1 3/4 × 1 3/4	3/16	7/16	2.1	5.0	0.63	1.50	0.51	0.64	0.18	0.48	0.14	0.35	0.54	0.56
1 1/2 × 1 1/2	3/16	3/8	1.8	3.6	0.53	1.09	0.44	0.55	0.11	0.25	0.10	0.22	0.46	0.48
1 1/4 × 1 1/4	1/8	1/4	1.0	2.0	0.30	0.61	0.35	0.43	0.044	0.098	0.05	0.10	0.38	0.40
1 1/8 × 1 1/8	1/8	1/4	0.9	1.8	0.27	0.55	0.32	0.39	0.032	0.071	0.04	0.08	0.34	0.36
1 × 1	1/8	3/16	0.8	1.2	0.23	0.36	0.30	0.33	0.022	0.035	0.03	0.05	0.30	0.31

ANGLES WITH UNEQUAL LEGS.

6 × 4	1⅝—1¾	13.9-26.4	4.18-7.93	1.96-2.17 0.96-1.17	15.5 -30.7 5.61-11.5	3.8 -7.3 1.8 -3.6	1.9 -2.0 1.2 -1.2
5 × 4	⅜—¾	10.8-22.0	3.23-6.61	1.53-1.74 1.03-1.24	8.16-17.5 4.67-10.3	2.4 -4.8 1.6 -3.3	1.6 -1.6 1.2 -1.2
5 × 3½	⅜—¾	10.2-20.3	3.05-6.23	1.61-1.82 0.86-1.07	7.78-16.7 3.18- 7.09	2.3 -4.7 1.2 -2.5	1.6 -1.6 1.0 -1.1
5 × 3	⅜—¾	9.5-19.5	2.86-5.86	1.70-1.91 0.70-0.91	7.37-15.87 2.04- 4.66	2.2 -4.6 0.89-1.9	1.6 -1.6 0.84-0.89
4 × 3½	⅜—¾	8.9-18.3	2.67-5.48	1.20-1.41 0.96-1.16	4.18- 9.14 2.99- 6.65	1.49-3.1 1.18-2.5	1.25-1.29 1.06-1.10
4 × 3	⅜—¾	8.3-17.0	2.48-5.11	1.28-1.49 0.78-0.99	3.96- 8.70 1.92- 4.38	1.46-3.0 0.87-1.8	1.26-1.30 0.88-0.93
3½ × 3	⅜—¾	7.7-15.8	2.30-4.73	1.08-1.29 0.83-1.04	2.72- 6.07 1.85- 4.21	1.13-2.4 0.85-1.8	1.09-1.13 0.90-0.84
3¼ × 2	¼—½	4.2- 8.5	1.25-2.56	1.10-1.24 0.48-0.61	1.36- 2.93 0.40- 0.91	0.63-1.3 0.26-0.56	1.04-1.07 0.57-0.60
3 × 2½	¼—½	4.4- 9.0	1.31-2.69	0.91-1.05 0.66-0.80	1.17- 2.54 0.74- 1.64	0.56-1.15 0.40-0.84	0.94-0.97 0.75-0.78
3 × 2	¼—½	4.0- 8.1	1.19-2.44	0.99-1.13 0.49-0.63	1.09- 2.36 0.39- 0.89	0.54-1.11 0.26-0.55	0.96-0.99 0.57-0.60
2½ × 2	¼—½	3.5- 7.3	1.06-2.19	0.79-0.92 0.54-0.67	0.65- 1.44 0.37- 0.85	0.38-0.79 0.25-0.54	0.78-0.81 0.59-0.62
2 × 1⅜	¼—⅜	2.6- 4.0	0.78-1.20	0.69-0.76 0.37-0.44	0.31- 0.50 0.12- 0.20	0.23-0.36 0.12-0.20	0.63-0.65 0.39-0.41

UNION IRON MILLS' ANGLE IRONS.

Weights per Foot corresponding to thicknesses varying by $\frac{1}{16}''$.

One cubic foot weighing 480 lbs.

Size. Inches.	$\frac{1}{8}''$	$\frac{3}{16}''$	$\frac{1}{4}''$	$\frac{5}{16}''$	$\frac{3}{8}''$	$\frac{7}{16}''$	$\frac{1}{2}''$	$\frac{9}{16}''$	$\frac{5}{8}''$	$\frac{11}{16}''$	$\frac{3}{4}''$	$\frac{13}{16}''$	$\frac{7}{8}''$
Equal Legs.													
6 × 6	..	..	..	..	..	..	19.2	21.7	24.2	26.7	29.2	31.7	34.2
4 × 4	..	..	..	..	9.5	11.2	12.9	14.5	16.2	17.9	19.5	..	..
3½ × 3½	..	..	..	..	8.3	9.7	11.2	12.7	14.1	15.6	17.0	..	..
3¼ × 3¼	..	..	..	..	7.7	9.0	10.4	11.7	13.1	14.4	15.8	..	..
3 × 3	..	..	..	5.9	7.2	8.4	9.7	10.9	12.2	..	..	..	..
2¾ × 2¾	..	..	..	5.4	6.5	7.7	8.8	..	..	..	..	..	..
2½ × 2½	..	..	..	4.9	5.9	7.0	8.0	..	..	..	..	..	..
2¼ × 2¼	..	..	3.5	4.5	5.4	6.4	7.3	..	..	..	..	..	..
2 × 2	..	..	3.1	4.0	4.8	5.6	..	..	..	..	..	..	..
1¾ × 1¾	..	2.1	2.8	3.5	4.3	5.0	..	..	..	..	..	..	..
1½ × 1½	..	1.8	2.4	3.0	3.6	..	..	..	..	..	..	..	..
1¼ × 1¼	1.0	1.5	2.0	..	..	..	..	..	..	..	..	..	..
1⅛ × 1⅛	0.9	1.4	1.8	..	..	..	..	..	..	..	..	..	..
1 × 1	0.8	1.2	1.6	..	..	..	..	..	..	..	..	..	..
¾ × ¾	0.6	0.9	..	..	..	..	..	..	..	..	..	..	..
Unequal Legs													
6 × 4	..	..	..	..	..	13.9	16.0	18.1	20.2	22.3	24.4	26.4	..
5 × 4	..	..	..	..	10.8	12.7	14.5	16.4	18.3	20.2	22.0	..	..
5 × 3½	..	..	..	..	10.2	11.9	13.7	15.5	17.2	19.0	20.8	..	..
5 × 3	..	..	..	..	9.5	11.2	12.9	14.5	16.2	17.9	19.5	..	..
4 × 3½	..	..	..	..	8.9	10.5	12.0	13.6	15.2	16.7	18.3	..	..
4 × 3	..	..	..	..	8.3	9.7	11.2	12.7	14.1	15.6	17.0	..	..
3½ × 3	..	..	..	..	7.7	9.0	10.4	11.7	13.1	14.4	15.8	..	..
3¼ × 2	..	..	4.2	5.3	6.4	7.4	8.5	..	..	..	..	..	..
3 × 2½	..	..	4.4	5.5	6.7	7.8	9.0	..	..	..	..	..	..
3 × 2	..	..	4.0	5.0	6.0	7.1	8.1	..	..	..	..	..	..
2½ × 2	..	..	3.5	4.5	5.4	6.4	7.3	..	..	..	..	..	..
2 × 1⅜	..	..	2.6	3.3	4.0	..	..	..	..	..	..	..	..

PROPERTIES OF UNION IRON MILLS'

T IRONS.

The moments of inertia and resistance, and radii of gyration, in this table, are close approximations only.

The table does not include all sizes manufactured.

Size, Flange by Stem. Inches.	Weight per Foot. Lbs.	Area of Section. Square Inches.	Distance of Center of Gravity from Top. Inches.	Moment of Inertia, neutral axis thro' center of gravity parallel to flange.	Least Moment of Resistance, neutral axis as before.	Radius of Gyration, neutral axis as before.	Moment of Inertia, neutral axis thro' center of gravity coincident with stem.	Least Moment of Resistance, neutral axis as before.	Radius of Gyration, neutral axis as before.
5 × 3	13	3.90	0.73	2.5	1.1	0.80	5.7	2.3	1.21
5 × 2½	10¼	3.08	0.58	1.4	0.71	0.66	4.6	1.8	1.21
4½ × 3½	15	4.50	1.13	5.2	2.18	1.07	3.9	1.7	0.93
4 × 5	14	4.20	1.57	10.5	3.05	1.57	2.7	1.4	0.80
4 × 4½	13½	4.05	1.37	7.8	2.48	1.39	2.7	1.4	0.82
4 × 4	12	3.60	1.18	5.4	1.91	1.22	2.6	1.3	0.84
4 × 3	9¼	2.78	0.80	2.1	0.96	0.87	2.3	1.1	0.90
4 × 2½	7½	2.25	0.62	1.1	0.60	0.70	2.0	1.0	0.93
4 × 2	6½	1.95	0.46	0.54	0.35	0.53	1.8	0.91	0.96
3½ × 4	11¼	3.38	1.24	5.15	1.87	1.23	1.8	1.00	0.72
3½ × 3½	10	3.00	1.04	3.34	1.36	1.05	1.6	0.93	0.73
3½ × 3	9¼	2.78	0.85	2.14	1.00	0.88	1.6	0.93	0.77
3 × 4	12¼	3.68	1.35	5.55	2.10	1.24	1.3	0.87	0.60
3 × 3½	11¾	3.53	1.15	3.93	1.67	1.06	1.4	0.92	0.62
3 × 3	7.6	2.28	0.90	1.89	0.90	0.91	0.94	0.63	0.64
3 × 2½	6	1.80	0.69	0.96	0.53	0.73	0.77	0.51	0.66
2½ × 3	6½	1.95	0.96	1.66	0.81	0.93	0.50	0.40	0.51
2½ × 2¾	6.6	1.98	0.86	1.39	0.74	0.84	0.55	0.44	0.53
2½ × 2½	5.4	1.62	0.75	0.91	0.43	0.75	0.46	0.37	0.53
2½ × 1¼	3	0.90	0.30	0.09	0.10	0.32	0.33	0.26	0.61

PROPERTIES OF UNION IRON MILLS'

STAR IRONS.

Size. Inches.	Weight per Foot. Lbs.	Thickness in Inches at End and Root of Flange.	Area. Sq. In.	Moment of Inertia, neutral axis thro' center of gravity.	Moment of Resistance, neutral axis as before.	Radius of Gyration, neutral axis as before.
4 × 4	12	3/8 — 9/16	3.60	2.32	1.16	0.81
3½ × 3½	9½	3/8 — 1/2	2.85	1.49	0.85	0.72
3 × 3	7¼	5/16 — 15/32	2.18	0.82	0.55	0.61
2½ × 2½	5½	5/16 — 13/32	1.65	0.45	0.36	0.52
2 × 2	3¾	1/4 — 13/32	1.13	0.20	0.20	0.43
1½ × 1½	2.3	3/16 — 5/16	0.69	0.065	0.087	0.31

EXPLANATION OF TABLE ON RIVETED GIRDERS.

Riveted girders are used in cases where rolled **I** Beams are insufficient to carry the load. On page 23 of the lithographed plates will be found illustrations of various forms of riveted girders. The sections with single webs are more economical than those with double webs (box girders), but the latter are stiffer laterally, and should always be used where the proportion of length of span to width of top flange is great and the girder is not held in position sideways. This proportion of length to width should not exceed twenty, without making provision for such increase by an addition of metal in the compression flange beyond that required by the table.

The web of the girder must be made of such thickness that there will be no tendency to buckle, and that the vertical shearing stress per square inch will not exceed 9000 lbs. This shearing stress is obtained by dividing half the load upon the girder by the web section. The first condition is attained when this shearing stress does not exceed $\dfrac{10000}{1 + \dfrac{d^2}{3000\,t^2}}$ in which d represents the depth of web of girder and t its thickness, both in inches. Ordinarily this formula gives a lower strain per square inch than 9000 lbs., so that both conditions are usually attained when the first is. Instead of increasing the thickness of the web, it may be stiffened also by means of vertical angle irons riveted to it at proper intervals. These should always be less than the depth of the girder, at least for the end panels, but towards the middle of the girder the stiffeners may be placed further apart or entirely omitted. Stiffeners should always be used at or near the supports, and at any other points where there is a concentration of heavy loads.

The rivets should be ¾″, unless the girder is light, when ⅝″ may be sufficient. The spacing ought not to exceed 6″ and should be closer for heavy flanges, but in all cases it should be close at the ends, say 3″ for a distance of 18″ to 24″ at each end.

The following table furnishes a ready means of determining the section of girder necessary to carry a certain load, for any span length from 10 to 39 feet, inclusive.

It will be noticed that the table is calculated for an allowed fiber strain of 10000 lbs. per square inch, while the tables on rolled beams are calculated for a fiber strain of 12000 lbs. per square inch. This reduction in the allowed strain is intended to cover the loss in strength, (somewhat greater than the loss in section,) due to the rivet holes, and the riveted girders proportioned by this table, will be found to be of about the same strength as the rolled beams, proportioned by the tables applying to them. The transverse strength of the web is neglected in the table.

The term flange, as applied to riveted girders, embraces all the metal in top or bottom of girder exclusive of web plate; or, in the case of a rolled beam or channel, with top and bottom plates, all the metal exclusive of web between fillets.

Girders intended to carry plastering, should be limited in depth, out to out, to $\frac{1}{24}$th of the span length or $\frac{1}{2}''$ per foot of this length, otherwise the deflection is liable to cause the plastering to crack.

EXAMPLE OF APPLICATION OF TABLE.

A $20''$ box girder is to carry a $13''$ brick wall equivalent to a weight of 30 tons over a space $20'$ in the clear. What size of girder is required?

Answer: The value of the coefficient for $20'$ span and $20''$ depth, as per table, $= 300$, and for $21'$ span and $20''$ depth $= 315$. The span, in this case, may be assumed at $20'$–$6''$, and the coefficient therefore at 307. Consequently $\dfrac{307 \times 30}{1000} = 9.21$ will be the area required in each flange. Making the top and bottom plates $12'' \times \frac{3}{8}''$, $= 4.5$ sq. in., there remain 4.7 sq. in. for the two angles, $= 8$ lbs. per foot apiece. Making the webs $20'' \times \frac{1}{4}''$, the shearing stress $= \dfrac{30 \times 2000 \times \frac{1}{2}}{2 \times 20 \times \frac{1}{4}} = 3000$ lbs. per square inch, which is also safe against buckling, since

$$\dfrac{10000}{1 + \dfrac{d^2}{3000\, t^2}} = \dfrac{10000}{1 + \dfrac{(20)^2}{3000\, (\frac{1}{4})^2}} = 3200 \text{ lbs., allowed.}$$

RIVETED GIRDERS.

Coefficients for determining the area required in flanges, allowing 10000 lbs. per square inch of gross section fiber strain:

Multiply the load, in tons of 2000 lbs., uniformly distributed, by the coefficient, and divide by 1000; the quotient will be the gross area, in square inches, required for each flange.

Distance between supports in Feet.	Depth of Girder, Out to Out of Web, in Inches.												
	12	14	16	18	20	22	24	26	28	30	32	34	36
10	250	214	188	167	150	136	125	115	107	100	94	88	83
11	275	236	206	183	165	150	138	127	118	110	103	97	92
12	300	257	225	200	180	164	150	138	129	120	113	106	100
13	325	279	244	217	195	177	163	150	139	130	122	115	108
14	350	300	263	233	210	191	175	162	150	140	131	124	117
15	375	321	281	250	225	205	188	173	161	150	141	132	125
16	400	343	300	267	240	218	200	185	171	160	150	141	133
17	425	364	319	283	255	232	213	196	182	170	159	150	142
18	450	386	338	300	270	245	225	208	193	180	169	159	150
19	475	407	356	317	285	259	238	219	204	190	178	168	158
20	500	429	375	333	300	273	250	231	214	200	188	176	167
21	525	450	394	350	315	286	263	242	225	210	197	185	175
22	550	471	413	367	330	300	275	254	236	220	206	194	183
23	575	493	431	383	345	314	288	265	246	230	216	203	192
24	600	514	450	400	360	327	300	277	257	240	225	212	200
25	625	536	469	417	375	341	313	288	268	250	234	221	208
26	650	557	488	433	390	355	325	300	279	260	244	229	217
27	675	579	506	450	405	368	338	312	289	270	253	238	225
28	700	600	525	467	420	382	350	323	300	280	263	247	233
29	725	621	544	483	435	395	363	335	311	290	272	256	242
30	750	643	563	500	450	409	375	346	321	300	281	265	250
31	775	664	581	517	465	423	388	358	332	310	291	274	258
32	800	686	600	533	480	436	400	369	343	320	300	282	267
33	825	707	619	550	495	450	413	381	354	330	309	291	275
34	850	729	638	567	510	464	425	392	364	340	319	300	283
35	875	750	656	583	525	477	438	404	375	350	328	309	292
36	900	771	675	600	540	491	450	415	386	360	338	318	300
37	925	793	694	617	555	505	463	427	396	370	347	326	308
38	950	814	713	633	570	518	475	438	407	380	356	335	317
39	975	836	731	650	585	532	488	450	418	390	366	344	325

COLUMNS AND STRUTS.

Explanation of tables, pages 77 to 81, inclusive.

The tables on Keystone Octagon and Piper's Patent Rivetless Columns give the areas and weights corresponding to different thicknesses of metal. Sections of these columns will be found on pages 13 and 14.

As it is impossible to repaint the inner surface of closed columns, or, at best, this is attended with much difficulty and expense, such columns should preferably be used only in the interior of buildings, where the changes in temperature are not considerable and the air is comparatively dry. In places exposed to the extremes of temperature and unprotected from the rain, the paint on the inner surface of the columns will, sooner or later, cease to be a protection to the iron from the moisture of the atmosphere, corrosion will set in, and, once begun, will continue as long as there is unoxidized metal left in the column.

Figures 1, 3 and 4, on page 26, represent types of columns with open sections, which admit of repainting at any time, and are therefore suitable for out-door work.

The table on the Ultimate Strength of Hollow Cylindrical Cast and Wrought Iron Columns gives the strains per square inch of section at which columns will fail, for various proportions of length of column to diameter.

To facilitate the use of the table, the length ($= l$) is expressed in feet, and the diameter ($= d$) in inches. The diameter to be assumed is the mean between the outside and inside diameters of the section.

Wrought iron columns fail either by deflecting bodily out of the straight line, or by the buckling of the metal between rivets or other points of support. Both actions may take place at the same time, but if the latter occurs by itself, it is an indication that the rivet spacing or the thickness of metal is insufficient; provided, however, that the length of column is greater than twelve diameters, as columns of shorter length fail generally by the buckling of the metal. The rule has been deduced from actual experiments, that the distance between centers of rivets in columns should not exceed, in the line of stress, sixteen times the thickness of metal of the parts joined, and that the distance between rivets or other points of support at right angles to the line of stress, should not exceed thirty times the thickness of metal.

The table on the Ultimate Strength of Wrought Iron Columns gives the strain per square inch of section at which columns will fail, for various proportions of length, in feet, to least radius of gyration, in inches. This table should be used for columns and struts which are not cylindrical, such as those represented by Figures 1, 2, 3, 4 and 5, on page 26.

If the column or strut is a single rolled beam, channel or other shape, the radius of gyration will be found in the foregoing tables on the properties of Union Iron Mills' Beams, Channels, etc.

If the column is composed of two channels latticed, as represented by Fig. 1, on page 26, the channels are usually placed far enough apart so that the column will be weakest in the direction of the webs, *i. e.*, with neutral axis at right angles to the webs; for which case the radius of gyration of the column section is the same as that of the single channel. But if the radius of gyration is wanted for the neutral axis through center of section parallel with web, obtain first the distance between center of gravity of channel and center of section, by the help of column 15 in table on the properties of Union Iron Mills' Channel Bars; the square root of this distance will be the radius of gyration of the section.

For a column section consisting of two channels with a beam between them, as in Fig. 3, on page 26, it is necessary to obtain first the moment of inertia of the section, whence the radius of gyration is found as the square root of the quotient of the moment of inertia divided by the area of the section. This moment of inertia, for a neutral axis through center of beam coincident with web, is equal to the sum of the moments of inertia of the beam and channels, as per tables on the properties of these shapes. The moment of inertia with neutral axis through center at right angles to web of beam, is found by adding the moment of inertia of the beam for this position of the axis, as per tables, to the product of the area of both channels multiplied by the square of the distance of the center of gravity of the channel from the center of the section. The moment of inertia, thus obtained, is approximate, being too small by the value of the moment of inertia of the channels with reference to a neutral axis through their centers of gravity parallel to the web, but the error is small and on the safe side.

For a section composed of three beams, as represented by Fig. 4, page 26, the correction for this approximation can be made, since the moments of inertia of beams with reference to an axis through their centers of gravity parallel to (coincident with) web is given in table for beams. In all other respects, proceed for this form of section as in the previous case.

If two channels are connected by means of two plates instead of a beam, as shown by Fig. 2, on page 26, the moment of inertia of the plates must be obtained instead of the beam. This moment of inertia, for a neutral axis through center of section perpendicular to the plates, is equal to the cube of the width of plate multiplied by $\frac{1}{12}$th of the thicknesses of the two plates added; and for a neutral axis parallel to plates, is equal to the area of the two plates multiplied by the square of the distance between the center of the plate and the center of the section.

A column is *square bearing* when it has square ends which butt against or are firmly connected with an immovable surface, such as the floor of a building; it is *pin and square bearing* when one end only is square bearing and the other presses against a close fitting pin, and it is *pin bearing* when both ends are thus pin-jointed, (for example, the posts in pin-connected bridges.)

With regard to the table on the Safe Resistance of Wooden Pillars, it should be said that comprehensive tests establishing the constants to be used in the formula have not been made to date, but it is believed that the values given in table err on the side of safety.

EXAMPLES OF APPLICATION OF TABLES.

I. What is the ultimate strength of a square bearing 10″ octagon column, ½″ thick and 20′ long?

Answer: The area of a 10″ × ½″ column, as per table on page 77, is 21.3 square inches. The mean diameter is 10″, very nearly, so that $\dfrac{1}{d} = \dfrac{20}{10} = 2$, for which the ultimate strength, as per table on page 79, = 33560 lbs. per square inch. Consequently the ultimate strength of the column = 33560 lbs. × 21.3 = 714800 lbs. The *safe* resistance for quiescent loads would be = ¼ × 714800 = 178700 lbs., and for moving loads = ⅕ × 714800 = 143000 lbs.

II. Required the ultimate strength of a 30 lb. 10″ beam used in the form of a strut, riveted at its ends so as to be firmly fixed, and measuring 10′ between the points where it is held in position.

By reference to table on page 64, the least radius of gyration of a 30 lb. 10″ beam is found to be = 0.94, (neutral axis coincident with web,) so that $\dfrac{1}{r} = \dfrac{10}{0.94} = 10.6$, for which the ultimate strength, as per table on page 80, = 27600 lbs. per square inch. The area of the beam being = 9 square inches, its ultimate strength will, therefore, = 9 × 27600 = 248400 lbs.

III. What is the radius of gyration of a column section composed of two 9″, 18 lb. channels, and a 6″, 13½ lb. beam, riveted together in the manner shown by Fig. 3, on page 26?

Answer, if neutral axis coincident with web of beam:

Moment of inertia of beam = 2.0
" " " " channels = 129.6
" " . " " section = 131.6

Area of section = 14.85 square inches. Therefore radius of gyration = $\sqrt{\dfrac{131.6}{14.85}} = 2.98$.

Answer, if neutral axis at right angles to web of beam:

Moment of inertia of beam = - - - . - 24.5.
Moment of inertia of channels = area of channels × distance of center of gravity from center of section squared = 10.8 square inches × 3.68^2 = - - 146.3
Moment of inertia of section = - - - - 170.8

Therefore radius of gyration = $\sqrt{\dfrac{170.8}{14.85}} = 3.39$.

KEYSTONE OCTAGON COLUMNS.

Thicknesses and Corresponding Areas and Weights per Foot.

Thickness Inch.	10 Inch Column.			8 Inch Column.			6 Inch Column.			4 Inch Column.			Thickness Inch.
	4 Segments		Weight of One Segment Lbs.	4 Segments		Weight of One Segment Lbs.	4 Segments		Weight of One Segment Lbs.	4 Segments		Weight of One Segment Lbs.	
	Area Sq. In.	Weight Lbs.		Area Sq. In.	Weight Lbs.		Area Sq. In.	Weight Lbs.		Area Sq. In.	Weight Lbs.		
3/16	....	....	....	....	....	....	5.60	18.7	4.7	3.91	13.0	3.3	3/16
1/4	....	....	....	9.78	32.6	8.2	7.13	23.8	5.9	4.98	16.6	4.2	1/4
5/16	14.22	47.4	11.9	11.80	39.3	9.8	8.66	28.9	7.2	6.05	20.2	5.0	5/16
3/8	16.58	55.3	13.8	13.81	46.0	11.5	10.20	34.0	8.5	7.12	23.7	5.9	3/8
7/16	18.94	63.1	15.8	15.83	52.8	13.2	11.73	39.1	9.8	8.20	27.3	6.8	7/16
1/2	21.30	71.0	17.8	17.85	59.5	14.9	13.26	44.2	11.1	9.27	30.9	7.7	1/2
9/16	23.66	78.9	19.7	19.86	66.2	16.6	14.79	49.3	12.3	....	....	....	9/16
5/8	26.01	86.7	21.7	21.88	72.9	18.2	16.32	54.4	13.6	....	....	....	5/8
11/16	28.37	94.6	23.6	23.89	79.6	19.9							11/16
3/4	30.73	102.4	25.6	25.91	86.4	21.6							3/4
13/16	33.09	110.3	27.6										13/16
7/8	35.45	118.2	29.5										7/8

PIPER'S PATENT RIVETLESS COLUMNS.

Thicknesses and Corresponding Areas and Weights per Foot.

Thickness Inch.	10 Inch Column.				8 Inch Column.				6 Inch Column.				4 Inch Column.			
	4 Segments, incl. Battens. Area. Sq. In.	Weight. Lbs.	Weight of One Segment Lbs.	Weight of One Batten. Lbs.	4 Segments, incl. Battens. Area. Sq. In.	Weight. Lbs.	Weight of One Segment Lbs.	Weight of One Batten. Lbs.	4 Segments, incl. Battens. Area. Sq. In.	Weight. Lbs.	Weight of One Segment Lbs.	Weight of One Batten. Lbs.	4 Segments, incl. Battens. Area. Sq. In.	Weight. Lbs.	Weight of One Segment Lbs.	Weight of One Batten. Lbs.
3/16	....	....	....		....	....	....		....	....	....		5.21	17.4	2.5	
1/4	....	....	....		10.98	36.6	6.1		7.30	24.3	4.2	1.87	6.00	20.0	3.1	
5/16	16.00	53.3	9.3		12.50	41.7	7.3		8.43	28.1	5.2		6.80	22.7	3.8	1.87
3/8	17.90	59.7	10.9		14.03	46.8	8.6		9.55	31.8	6.1		7.60	25.3	4.5	
7/16	19.80	66.0	12.5		15.55	51.8	9.9	3.1	10.68	35.6	7.0	2.3	8.39	28.0	5.1	
1/2	21.70	72.3	14.1		17.08	56.9	11.1		11.81	39.4	8.0					
9/16	23.60	78.7	15.7	4.0	18.60	62.0	12.4									
5/8	25.50	85.0	17.3		20.13	67.1	13.7									
11/16	27.40	91.3	18.8													
3/4	29.30	97.7	20.4													

ULTIMATE STRENGTH OF
HOLLOW CYLINDRICAL CAST AND WROUGHT IRON COLUMNS,

For different proportions of length in feet ($= l$)
To least diameter in inches ($= d$).
Ultimate Strength in lbs. per square inch =

Cast Iron.			Wrought Iron.		
Square Bearing:	Pin & Square:	Pin Bearing:	Square Bearing:	Pin & Square:	Pin Bearing:
80000	80000	80000	40000	40000	40000
$1+\dfrac{(12\,l)^2}{800\,d^2}$	$1+\dfrac{3(12\,l)^2}{1600\,d^2}$	$1+\dfrac{(12\,l)^2}{400\,d^2}$	$1+\dfrac{(12\,l)^2}{3000\,d^2}$	$1+\dfrac{(12\,l)^2}{2000\,d^2}$	$1+\dfrac{(12\,l)^2}{1500\,d^2}$

To obtain Safe Resistance:

For quiescent loads (buildings) divide by $\begin{cases} 6 \text{ for cast iron,} \\ 4 \text{ for wrought iron.} \end{cases}$

For moving loads (bridges) divide by $\begin{cases} 8 \text{ for cast iron,} \\ 5 \text{ for wrought iron.} \end{cases}$

$\dfrac{l}{d}$	Cast Iron. Ultimate Strength in Lbs. per sq. in.			Wrought Iron. Ultimate Strength in Lbs. per sq. in.		
	Square.	Pin and Square.	Pin.	Square.	Pin and Square.	Pin.
1.0	67800	62990	58820	38170	37310	36500
1.1	65690	60300	55730	37800	36790	35840
1.2	63530	57600	52690	37410	36240	35140
1.3	61340	54930	49740	37000	35660	34420
1.4	59140	52310	46900	36560	35050	33670
1.5	56940	49770	44200	36100	34420	32890
1.6	54760	47300	41630	35620	33770	32110
1.7	52620	44940	39210	35130	33110	31320
1.8	50530	42670	36930	34620	32430	30510
1.9	48490	40510	34790	34090	31750	29710
2.0	46510	38460	32790	33560	31060	28900
2.1	44600	36520	30920	33010	30360	28100
2.2	42750	34680	29180	32460	29660	27310
2.3	40980	32940	27540	31900	28970	26530
2.4	39280	31310	26030	31340	28270	25760
2.5	37650	29770	24620	30770	27590	25000
2.6	36090	28320	23300	30200	26900	24260
2.7	34600	26950	22070	29630	26230	23530
2.8	33180	25670	20930	29060	25570	22820
2.9	31820	24460	19860	28500	24910	22130
3.0	30530	23320	18870	27930	24270	21460
3.1	29310	22250	17940	27370	23640	20810
3.2	28140	21250	17070	26820	23020	20170
3.3	27030	20300	16260	26270	22420	19560
3.4	25970	19410	15500	25730	21830	18960

ULTIMATE STRENGTH OF WROUGHT IRON COLUMNS,

For different proportions of length in feet $(= l)$
To least radius of gyration in inches $(= r)$.

Ultimate Strength in lbs. per square inch =

Column Square Bearing:	Column Pin and Square Bearing:	Column Pin Bearing:
$\dfrac{40000}{1 + \dfrac{(12\,l)^2}{36000\,r^2}}$	$\dfrac{40000}{1 + \dfrac{(12\,l)^2}{24000\,r^2}}$	$\dfrac{40000}{1 + \dfrac{(12\,l)^2}{18000\,r^2}}$

To obtain Safe Resistance:
For quiescent loads, as in buildings, divide by 4.
For moving loads, as in bridges, divide by 5.

$\dfrac{l}{r}$	Ultimate Strength in Lbs. per square inch.			$\dfrac{l}{r}$	Ultimate Strength in Lbs. per square inch.		
	Square.	Pin and Square.	Pin.		Square.	Pin and Square.	Pin.
3.0	38610	37950	37310	8.0	31850	28900	26460
3.2	38430	37680	36970	8.2	31520	28500	26010
3.4	38230	37400	36610	8.4	31190	28100	25570
3.6	38030	37110	36240	8.6	30870	27700	25130
3.8	37820	36810	35860	8.8	30540	27310	24700
4.0	37590	36500	35460	9.0	30210	26920	24270
4.2	37360	36170	35050	9.2	29880	26530	23850
4.4	37120	35840	34640	9.4	29550	26140	23430
4.6	36870	35500	34210	9.6	29230	25760	23030
4.8	36620	35140	33770	9.8	28900	25370	22620
5.0	36360	34780	33330	10.0	28570	25000	22220
5.2	36090	34420	32890	10.2	28250	24630	21830
5.4	35820	34050	32440	10.4	27920	24260	21440
5.6	35540	33670	31980	10.6	27600	23890	21060
5.8	35260	33280	31520	10.8	27270	23530	20690
6.0	34970	32890	31060	11.0	26950	23170	20330
6.2	34670	32500	30590	11.2	26640	22820	19960
6.4	34370	32110	30130	11.4	26320	22470	19610
6.6	34060	31710	29670	11.6	26000	22130	19270
6.8	33750	31310	29200	11.8	25690	21800	18930
7.0	33440	30910	28740	12.0	25380	21460	18590
7.2	33130	30510	28270	12.2	25070	21130	18260
7.4	32810	30110	27820	12.4	24770	20810	17940
7.6	32490	29710	27360	12.6	24470	20490	17620
7.8	32170	29310	26910	12.8	24170	20180	17310

ULTIMATE STRENGTH OF
RECTANGULAR TIMBER PILLARS, WELL SEASONED,

For different proportions of length in feet ($= l$)
To least diameter or side in inches ($= d$).

Ultimate Strength in lbs. per square inch =

$$\text{Pillar Square Bearing:} \quad \frac{5600}{1 + \frac{(12\,l)^2}{550\,d^2}}$$

$$\text{Pillar Pin and Square Bearing:} \quad \frac{5600}{1 + \frac{1.5\,(12\,l)^2}{550\,d^2}}$$

$$\text{Pillar Pin Bearing:} \quad \frac{5600}{1 + \frac{(12\,l)^2}{275}}$$

The above formula for Square Bearing Pillars is based upon Lemande's experiments on French oak, and agrees fairly with Hodgkinson's formula for French oak pillars of 30 diameters and over.

The strength of pillars of French oak, Red deal and Dantzig oak, is given by Hodgkinson as proportional to the ratio, 6.9 : 7.8 : 10.95.

It is believed the above formulæ for French oak and the following table calculated from them, will also apply to American white pine of best quality.

Green timber has only about half the strength of dry.

To obtain the Safe Resistance, divide by 6.

$\dfrac{l}{d}$	Ultimate Strength in Lbs. per square inch.			$\dfrac{l}{d}$	Ultimate Strength in Lbs. per square inch.		
	Square.	Pin and Square.	Pin.		Square.	Pin and Square.	Pin.
1.0	4440	4020	3680	2.5	2120	1620	1310
1.1	4250	3800	3430	2.6	2020	1530	1230
1.2	4070	3580	3190	2.7	1930	1450	1160
1.3	3880	3370	2970	2.8	1830	1370	1100
1.4	3700	3160	2760	2.9	1750	1300	1040
1.5	3520	2970	2570	3.0	1670	1230	980
1.6	3350	2790	2390	3.1	1590	1170	930
1.7	3190	2620	2230	3.2	1520	1120	880
1.8	3040	2470	2080	3.3	1450	1060	840
1.9	2890	2320	1940	3.4	1390	1010	790
2.0	2740	2180	1810	3.5	1330	960	760
2.1	2600	2050	1690	3.6	1270	920	720
2.2	2470	1930	1580	3.7	1220	880	690
2.3	2350	1820	1490	3.8	1170	840	650
2.4	2230	1720	1400	3.9	1120	800	620

GENERAL NOTES ON FLOORS and ROOFS.

On page 23 will be found examples of floor joists and their connections. When two beams are placed side by side, as in Fig. 1, they should be connected together by means of bolts and cast-iron separators, fitted closely between the flanges of the beams. The office of these separators is to hold in position the compression flange of the beams, preventing side deflection or buckling, and to firmly unite the two beams, so that they will act in unison. Separators should be used near the supports and at distances of five or six feet. They are shown by Figs. 2 and 3, on page 24. Their weight will range from 19 lbs. for the heavy 15″ beams, to 5 lbs. for 6″ beams.

Figures 1, 2 and 3 show the methods of connecting beams with each other. In Figs. 1 and 2 the lighter beam is coped into the heavier one, the weight being carried by the lower flange of the latter. The angle with which the webs are connected, serves only to hold the beams in position, in this case. In Fig. 3 the load of the smaller beams is transferred to the larger by means of angles riveted to the webs, and in case this is not sufficient, an angle may be riveted to the web of the larger beam underneath the smaller, as shown, to assist in carrying the load.

Figures 5, 6, 7, 8, 9 and 10, on page 23, are illustrations of various forms of girders, such as it is often necessary to use in the front of buildings, to carry walls, or in the interior, to support the joists. Where these girders rest upon the wall, cast or wrought-iron bed plates should be used, to distribute the pressure over a greater surface, and thereby prevent the crushing of the brick directly under the girder. In some cases a tough, large size stone will answer without the plates, but where the pressure is heavy, both plates and stone should be used. Figs. 5, 6, 9 and 10, are illustrations.

On page 24, Fig. 1, is represented a girder composed of two beams, carrying a brick wall, in position. In case of failure of the girder, only a part of the wall above it would drop down, the line of rupture for brick-work making an angle of about 30° with the vertical, called the angle of repose. The weight to be carried by the girder may, therefore, be considered to be only that part of the wall between the lines of rupture, provided, that in building the wall, the center of the girder was supported temporarily with a wooden prop, preventing deflection. Several courses should, however, be laid before this is done.

If $l =$ the clear span of girder, and $h =$ the hight of wall above it, the superficial area of the trapezoid between the lines of rupture, is expressed by $h(2l - 1.2h)$, but deductions must, of course, be made for windows or other openings in the wall, if there are any.

In order to be entirely on the safe side, and also for the sake of simplicity, the weight of wall between vertical lines directly over the girder, is frequently adopted as the load to be carried by it.

Weight of Brick-work per Superficial Foot, for a

9″ wall = 84 lbs., 13″ wall = 121 lbs., 18″ wall = 168 lbs., one cubic foot weighing 112 lbs.

There are various fire-proof floors in use; one of the most common is that represented by Fig. 1, on page 23. Four-inch brick arches are built between beams spaced *not over* 5 feet apart, and tied together by rods ¾″ to 1″ diameter, at intervals of 4′ to 6′, so as to take the thrust of the arches off the walls. Tee or angle irons are inserted in the wall, so as to hold it firmly in line between the points held by the rods. The top of the arches is leveled off with concrete, allowing space, however, for wooden strips, to which the floor timber is nailed. The plastering for ceiling usually covers the arches only, so that the ceiling will appear curved and show the lower flanges of the iron beams.

A convenient device for centering the arches is shown in Fig. 4. The centers are suspended by iron hooks from the lower flanges of the beams, and can be moved forward and back, and removed at pleasure.

Figure 4, on page 24, and Fig. 3, on page 25, are examples of flush, plastered ceilings, the laths in the latter case being held by light castings. Fig. 3, on page 24, is an example of an iron ceiling, composed of sheet iron pressed to suitable form, laid upon the lower flanges of the beams; and Figs. 2 and 5 are illustrations of corrugated iron ceilings. Both are open to the objection that the condensed moisture of the air will collect upon the iron and fall into the rooms below. Particularly is this the case in rooms filled with people, and such ceilings should, therefore, be restricted in their use, or the iron should be covered in such manner from below, that the access of the air is effectually cut off, as by plastering.

The weight of a fire-proof floor, consisting of four-inch brick arches between beams, with concrete filling above the arches and flooring, will generally average about 70 lbs. per square foot, exclusive of the weight of the beams. The following are average weights of some other constructions, and the usual assumptions made for superimposed load:

Iron roof of 100 feet span, with corrugated iron laid directly upon purlins, will weigh

Approximately,	10 lbs. ⌽ sq. ft.	
If boarded, add	3 "	"
For lathed and plastered ceiling, allow	10 "	"
For snow and vertical component of wind force, allow	30 "	"
For superimposed load on		
Floors of dwellings, assume	60 "	"
" " churches, theaters and ball rooms,	125 "	"
" " warehouses,	250 "	"
Weight of snow, freshly fallen,	5 to 12 "	cub. ft.
" " saturated, (slush,)	40 "	"
Crowd of people, closely packed,	80 "	sq. ft.
Wind pressure (violent hurricane,)	50 "	"

Rule for finding the sectional area of a bar of wrought iron, given the weight per foot:

Multiply by 3 and divide by 10.

Rule for finding the weight per foot, given the area:

Divide by 3 and multiply by 10.

CORRUGATED AND GALVANIZED IRON.

Corrugated Iron is used for roofs and sides of buildings. It is usually laid directly upon the purlins in roofs, and held in place by means of clips of hoop iron, which encircle the purlin and are placed in distances of about twelve inches apart. Special care must be taken that the projecting edges of the corrugated iron, at the eaves and gable ends of the roof, are well secured, otherwise the wind will loosen the sheets and fold them up.

The corrugations are made of various sizes; the smaller present a more pleasing appearance to the eye, while the larger are stiffer and will span a greater distance, thereby permitting the purlins to be placed further apart. The sizes of sheets generally used for both roofing and siding, are No. 20 and 22.

The corrugated iron which will be described in the following, is manufactured by the Keystone Bridge Company, of Pittsburgh. It is of medium size, presenting both a good appearance and being of sufficient strength for usual requirements.

By one corrugation is meant the double curve between corresponding points, and by depth of corrugation, the greatest deviation from the straight line, measured between the concave surfaces of the corrugated sheet.

The Keystone Bridge Company's corrugations are $2.425''$ long, measured on the straight line; they require a length of iron of $2.725''$ to make one corrugation, and the depth of corrugation is $\frac{1}{2}''$. One corrugation is allowed for lap in the width of the sheet and $6''$ in the length, for the usual pitch of roof of two to one. Sheets can be corrugated of any length not exceeding ten feet. The most advantageous width is $30\frac{1}{2}''$, which (allowing $\frac{1}{2}''$ for irregularities) will make eleven corrugations $= 30''$, or, making allowance for laps, will cover $24\frac{1}{4}''$ of the surface of the roof.

By actual trial it was found that corrugated iron No. 20, spanning 6 feet, will begin to give a permanent deflection for a load of 30 lbs. per square foot, and that it will collapse with a load of 60 lbs. per square foot. The distance between centers of purlins should therefore not exceed 6 feet, and, preferably, be less than this.

KEYSTONE BRIDGE CO.'S CORRUGATED IRON.

The following table is calculated for sheets 30½″ wide before corrugating.

No. by Birmingham Gauge.	Thickness. Inch.	Weight per Square Foot, Flat. Lbs.	Weight per Square Foot, Corrugated. Lbs.	Weight per Square of 100 square feet, when laid, allowing 6″ lap in length and 2½″ or one corrugation in width of sheet, for sheet lengths of:						Weight per Square Foot, Flat, Galvanized. Lbs.
				5′	6′	7′	8′	9′	10′	
16	.065	2.61	3.28	365	358	353	350	348	346	2.95
18	.049	1.97	2.48	275	270	267	264	262	261	2.31
20	.035	1.40	1.76	196	192	190	188	186	185	1.74
22	.028	1.12	1.41	156	154	152	150	149	148	1.46
24	.022	.88	1.11	123	121	119	118	117	117	1.22
26	.018	.72	.91	101	99	97	97	96	95	1.06

RESULTS OF TEST

of a corrugated sheet No. 20, 2′–0″ wide, 6′–0″ long between supports, loaded uniformly with fire clay.

Load per Square Foot. Lbs.	Deflection at Center under Load. Inches.	Permanent Deflection, Load Removed.
5	½	0
10	¾	0
15	1	0
20	1¼	0
25	1½	0
30	1⅞	⅛
35	2¼	½
40	2⅝	¾
45	3½	1⅛
50	4	1½
55	6½	Not Noted.
60	Broke Down.	" "

ILLUSTRATION OF APPLICATION

OF TABLES ON FLAT ROLLED IRON.

Pages 88 to 99, inclusive.

What is the weight per foot of a bar $5'' \times 1\frac{1}{16}''$ in section? *Answer:* In the column for $5''$ width, and in the line for $1\frac{1}{16}''$ thickness, will be found the value 17.71, which is the weight desired.

What thickness of $4\frac{1}{2}''$ bar will be required to give an area of 5.3 square inches? *Answer:* In the column for $4\frac{1}{2}''$ width will be found 5.34, which is the area nearest to that required; the corresponding thickness being $1\frac{3}{16}''$, the bar should be $4\frac{1}{2}'' \times 1\frac{3}{16}''$.

ILLUSTRATION OF APPLICATION

OF TABLES ON DECIMAL PARTS OF A FOOT FOR EACH $\frac{1}{64}$th OF AN INCH.

Pages 100 to 103, inclusive.

What is the value of $5' - 7\frac{11}{64}''$, expressed in feet and decimals of a foot? *Answer:* 5.5977; found by looking in column for $7''$, and in line for $\frac{11}{64}''$.

What is the value of 11.6838′, expressed in feet, inches and fractions of an inch? *Answer:* The value nearest to the decimal .6838, to be found in table, is .6836, which is $= 8\frac{13}{64}''$, therefore 11.6838′ $= 11' - 8\frac{13}{64}''$, nearly.

WEIGHTS OF FLAT ROLLED IRON PER LINEAL FOOT.

For Thicknesses from 1/16 in. to 2 in. and Widths from 1 in. to 12¾ in.
Iron weighing 480 lbs. per cubic foot.

Thickness in Inches.	1″	1¼″	1½″	1¾″	2″	2¼″	2½″	2¾″	12″
1/16	.208	.260	.313	.365	.417	.469	.521	.573	2.50
1/8	.417	.521	.625	.729	.833	.938	1.04	1.15	5.00
3/16	.625	.781	.938	1.09	1.25	1.41	1.56	1.72	7.50
¼	.833	1.04	1.25	1.46	1.67	1.88	2.08	2.29	10.00
5/16	1.04	1.30	1.56	1.82	2.08	2.34	2.60	2.86	12.50
3/8	1.25	1.56	1.88	2.19	2.50	2.81	3.13	3.44	15.00
7/16	1.46	1.82	2.19	2.55	2.92	3.28	3.65	4.01	17.50
½	1.67	2.08	2.50	2.92	3.33	3.75	4.17	4.58	20.00
9/16	1.88	2.34	2.81	3.28	3.75	4.22	4.69	5.16	22.50
5/8	2.08	2.60	3.13	3.65	4.17	4.69	5.21	5.73	25.00
11/16	2.29	2.86	3.44	4.01	4.58	5.16	5.73	6.30	27.50
¾	2.50	3.13	3.75	4.38	5.00	5.63	6.25	6.88	30.00
13/16	2.71	3.39	4.06	4.74	5.42	6.09	6.77	7.45	32.50
7/8	2.92	3.65	4.38	5.10	5.83	6.56	7.29	8.02	35.00
15/16	3.13	3.91	4.69	5.47	6.25	7.03	7.81	8.59	37.50
1	3.33	4.17	5.00	5.83	6.67	7.50	8.33	9.17	40.00
1 1/16	3.54	4.43	5.31	6.20	7.08	7.97	8.85	9.74	42.50
1 1/8	3.75	4.69	5.63	6.56	7.50	8.44	9.38	10.31	45.00
1 3/16	3.96	4.95	5.94	6.93	7.92	8.91	9.90	10.89	47.50
1 ¼	4.17	5.21	6.25	7.29	8.33	9.38	10.42	11.46	50.00
1 5/16	4.37	5.47	6.56	7.66	8.75	9.84	10.94	12.03	52.50
1 3/8	4.58	5.73	6.88	8.02	9.17	10.31	11.46	12.60	55.00
1 7/16	4.79	5.99	7.19	8.39	9.58	10.78	11.98	13.18	57.50
1 ½	5.00	6.25	7.50	8.75	10.00	11.25	12.50	13.75	60.00
1 9/16	5.21	6.51	7.81	9.11	10.42	11.72	13.02	14.32	62.50
1 5/8	5.42	6.77	8.13	9.48	10.83	12.19	13.54	14.90	65.00
1 11/16	5.63	7.03	8.44	9.84	11.25	12.66	14.06	15.47	67.50
1 ¾	5.83	7.29	8.75	10.21	11.67	13.13	14.58	16.04	70.00
1 13/16	6.04	7.55	9.06	10.57	12.08	13.59	15.10	16.61	72.50
1 7/8	6.25	7.81	9.38	10.94	12.50	14.06	15.63	17.19	75.00
1 15/16	6.46	8.07	9.69	11.30	12.92	14.53	16.15	17.76	77.50
2	6.67	8.33	10.00	11.67	13.33	15.00	16.67	18.33	80.00

WEIGHTS OF FLAT ROLLED IRON PER LINEAL FOOT.

(CONTINUED.)

Thickness in Inches.	3″	3¼″	3½″	3¾″	4″	4¼″	4½″	4¾″	12″
1/16	.625	.677	.729	.781	.833	.885	.938	.990	2.50
1/8	1.25	1.35	1.46	1.56	1.67	1.77	1.88	1.98	5.00
3/16	1.88	2.03	2.19	2.34	2.50	2.66	2.81	2.97	7.50
1/4	2.50	2.71	2.92	3.13	3.33	3.54	3.75	3.96	10.00
5/16	3.13	3.39	3.65	3.91	4.17	4.43	4.69	4.95	12.50
3/8	3.75	4.06	4.38	4.69	5.00	5.31	5.63	5.94	15.00
7/16	4.38	4.74	5.10	5.47	5.83	6.20	6.56	6.93	17.50
1/2	5.00	5.42	5.83	6.25	6.67	7.08	7.50	7.92	20.00
9/16	5.63	6.09	6.56	7.03	7.50	7.97	8.44	8.91	22.50
5/8	6.25	6.77	7.29	7.81	8.33	8.85	9.38	9.90	25.00
11/16	6.88	7.45	8.02	8.59	9.17	9.74	10.31	10.89	27.50
3/4	7.50	8.13	8.75	9.38	10.00	10.63	11.25	11.88	30.00
13/16	8.13	8.80	9.48	10.16	10.83	11.51	12.19	12.86	32.50
7/8	8.75	9.48	10.21	10.94	11.67	12.40	13.13	13.85	35.00
15/16	9.38	10.16	10.94	11.72	12.50	13.28	14.06	14.84	37.50
1	10.00	10.83	11.67	12.50	13.33	14.17	15.00	15.83	40.00
1 1/16	10.63	11.51	12.40	13.28	14.17	15.05	15.94	16.82	42.50
1 1/8	11.25	12.19	13.13	14.06	15.00	15.94	16.88	17.81	45.00
1 3/16	11.88	12.86	13.85	14.84	15.83	16.82	17.81	18.80	47.50
1 1/4	12.50	13.54	14.58	15.63	16.67	17.71	18.75	19.79	50.00
1 5/16	13.13	14.22	15.31	16.41	17.50	18.59	19.69	20.78	52.50
1 3/8	13.75	14.90	16.04	17.19	18.33	19.48	20.63	21.77	55.00
1 7/16	14.38	15.57	16.77	17.97	19.17	20.36	21.56	22.76	57.50
1 1/2	15.00	16.25	17.50	18.75	20.00	21.25	22.50	23.75	60.00
1 9/16	15.63	16.93	18.23	19.53	20.83	22.14	23.44	24.74	62.50
1 5/8	16.25	17.60	18.96	20.31	21.67	23.02	24.38	25.73	65.00
1 11/16	16.88	18.28	19.69	21.09	22.50	23.91	25.31	26.72	67.50
1 3/4	17.50	18.96	20.42	21.88	23.33	24.79	26.25	27.71	70.00
1 13/16	18.13	19.64	21.15	22.66	24.17	25.68	27.19	28.70	72.50
1 7/8	18.75	20.31	21.88	23.44	25.00	26.56	28.13	29.69	75.00
1 15/16	19.38	20.99	22.60	24.22	25.83	27.45	29.06	30.68	77.50
2	20.00	21.67	23.33	25.00	26.67	28.33	30.00	31.67	80.00

WEIGHTS OF FLAT ROLLED IRON PER LINEAL FOOT.

(CONTINUED.)

Thickness in Inches.	5″	5¼″	5½″	5¾″	6″	6¼″	6½″	6¾″	12″
1/16	1.04	1.09	1.15	1.20	1.25	1.30	1.35	1.41	2.50
1/8	2.08	2.19	2.29	2.40	2.50	2.60	2.71	2.81	5.00
3/16	3.13	3.28	3.44	3.59	3.75	3.91	4.06	4.22	7.50
1/4	4.17	4.38	4.58	4.79	5.00	5.21	5.42	5.63	10.00
5/16	5.21	5.47	5.73	5.99	6.25	6.51	6.77	7.03	12.50
3/8	6.25	6.56	6.88	7.19	7.50	7.81	8.13	8.44	15.00
7/16	7.29	7.66	8.02	8.39	8.75	9.11	9.48	9.84	17.50
1/2	8.33	8.75	9.17	9.58	10.00	10.42	10.83	11.25	20.00
9/16	9.38	9.84	10.31	10.78	11.25	11.72	12.19	12.66	22.50
5/8	10.42	10.94	11.46	11.98	12.50	13.02	13.54	14.06	25.00
11/16	11.46	12.03	12.60	13.18	13.75	14.32	14.90	15.47	27.50
3/4	12.50	13.13	13.75	14.38	15.00	15.63	16.25	16.88	30.00
13/16	13.54	14.22	14.90	15.57	16.25	16.93	17.60	18.28	32.50
7/8	14.58	15.31	16.04	16.77	17.50	18.23	18.96	19.69	35.00
15/16	15.63	16.41	17.19	17.97	18.75	19.53	20.31	21.09	37.50
1	16.67	17.50	18.33	19.17	20.00	20.83	21.67	22.50	40.00
1 1/16	17.71	18.59	19.48	20.36	21.25	22.14	23.02	23.91	42.50
1 1/8	18.75	19.69	20.63	21.56	22.50	23.44	24.38	25.31	45.00
1 3/16	19.79	20.78	21.77	22.76	23.75	24.74	25.73	26.72	47.50
1 1/4	20.83	21.88	22.92	23.96	25.00	26.04	27.08	28.13	50.00
1 5/16	21.88	22.97	24.06	25.16	26.25	27.34	28.44	29.53	52.50
1 3/8	22.92	24.06	25.21	26.35	27.50	28.65	29.79	30.94	55.00
1 7/16	23.96	25.16	26.35	27.55	28.75	29.95	31.15	32.34	57.50
1 1/2	25.00	26.25	27.50	28.75	30.00	31.25	32.50	33.75	60.00
1 9/16	26.04	27.34	28.65	29.95	31.25	32.55	33.85	35.16	62.50
1 5/8	27.08	28.44	29.79	31.15	32.50	33.85	35.21	36.56	65.00
1 11/16	28.13	29.53	30.94	32.34	33.75	35.16	36.56	37.97	67.50
1 3/4	29.17	30.63	32.08	33.54	35.00	36.46	37.92	39.38	70.00
1 13/16	30.21	31.72	33.23	34.74	36.25	37.76	39.27	40.78	72.50
1 7/8	31.25	32.81	34.38	35.94	37.50	39.06	40.63	42.19	75.00
1 15/16	32.29	33.91	35.52	37.14	38.75	40.36	41.98	43.59	77.50
2	33.33	35.00	36.67	38.33	40.00	41.67	43.33	45.00	80.00

WEIGHTS OF FLAT ROLLED IRON PER LINEAL FOOT.

(CONTINUED.)

Thickness in Inches.	7″	7¼″	7½″	7¾″	8″	8¼″	8½″	8¾″	12″
1/16	1.46	1.51	1.56	1.61	1.67	1.72	1.77	1.82	2.50
1/8	2.92	3.02	3.13	3.23	3.33	3.44	3.54	3.65	5.00
3/16	4.38	4.53	4.69	4.84	5.00	5.16	5.31	5.47	7.50
1/4	5.83	6.04	6.25	6.46	6.67	6.88	7.08	7.29	10.00
5/16	7.29	7.55	7.81	8.07	8.33	8.59	8.85	9.11	12.50
3/8	8.75	9.06	9.38	9.69	10.00	10.31	10.63	10.94	15.00
7/16	10.21	10.57	10.94	11.30	11.67	12.03	12.40	12.76	17.50
1/2	11.67	12.08	12.50	12.92	13.33	13.75	14.17	14.58	20.00
9/16	13.13	13.59	14.06	14.53	15.00	15.47	15.94	16.41	22.50
5/8	14.58	15.10	15.63	16.15	16.67	17.19	17.71	18.23	25.00
11/16	16.04	16.61	17.19	17.76	18.33	18.91	19.48	20.05	27.50
3/4	17.50	18.13	18.75	19.38	20.00	20.63	21.25	21.88	30.00
13/16	18.96	19.64	20.31	20.99	21.67	22.34	23.02	23.70	32.50
7/8	20.42	21.15	21.88	22.60	23.33	24.06	24.79	25.52	35.00
15/16	21.88	22.66	23.44	24.22	25.00	25.78	26.56	27.34	37.50
1	23.33	24.17	25.00	25.83	26.67	27.50	28.33	29.17	40.00
1 1/16	24.79	25.68	26.56	27.45	28.33	29.22	30.10	30.99	42.50
1 1/8	26.25	27.19	28.13	29.06	30.00	30.94	31.88	32.81	45.00
1 3/16	27.71	28.70	29.69	30.68	31.67	32.66	33.65	34.64	47.50
1 1/4	29.17	30.21	31.25	32.29	33.33	34.38	35.42	36.46	50.00
1 5/16	30.62	31.72	32.81	33.91	35.00	36.09	37.19	38.28	52.50
1 3/8	32.08	33.23	34.38	35.52	36.67	37.81	38.96	40.10	55.00
1 7/16	33.54	34.74	35.94	37.14	38.33	39.53	40.73	41.93	57.50
1 1/2	35.00	36.25	37.50	38.75	40.00	41.25	42.50	43.75	60.00
1 9/16	36.46	37.76	39.06	40.36	41.67	42.97	44.27	45.57	62.50
1 5/8	37.92	39.27	40.63	41.98	43.33	44.69	46.04	47.40	65.00
1 11/16	39.38	40.78	42.19	43.59	45.00	46.41	47.81	49.22	67.50
1 3/4	40.83	42.29	43.75	45.21	46.67	48.13	49.58	51.04	70.00
1 13/16	42.29	43.80	45.31	46.82	48.33	49.84	51.35	52.86	72.50
1 7/8	43.75	45.31	46.88	48.44	50.00	51.56	53.13	54.69	75.00
1 15/16	45.21	46.82	48.44	50.05	51.67	53.28	54.90	56.51	77.50
2	46.67	48.33	50.00	51.67	53.33	55.00	56.67	58.33	80.00

WEIGHTS OF FLAT ROLLED IRON PER LINEAL FOOT.

(CONTINUED.)

Thickness in Inches.	9″	9¼″	9½″	9¾″	10″	10¼″	10½″	10¾″	12″
1/16	1.88	1.93	1.98	2.03	2.08	2.14	2.19	2.24	2.50
1/8	3.75	3.85	3.96	4.06	4.17	4.27	4.38	4.48	5.00
3/16	5.63	5.78	5.94	6.09	6.25	6.41	6.56	6.72	7.50
1/4	7.50	7.71	7.92	8.13	8.33	8.54	8.75	8.96	10.00
5/16	9.38	9.64	9.90	10.16	10.42	10.68	10.94	11.20	12.50
3/8	11.25	11.56	11.88	12.19	12.50	12.81	13.13	13.44	15.00
7/16	13.13	13.49	13.85	14.22	14.58	14.95	15.31	15.68	17.50
1/2	15.00	15.42	15.83	16.25	16.67	17.08	17.50	17.92	20.00
9/16	16.88	17.34	17.81	18.28	18.75	19.22	19.69	20.16	22.50
5/8	18.75	19.27	19.79	20.31	20.83	21.35	21.88	22.40	25.00
11/16	20.63	21.20	21.77	22.34	22.92	23.49	24.06	24.64	27.50
3/4	22.50	23.13	23.75	24.38	25.00	25.62	26.25	26.88	30.00
13/16	24.38	25.05	25.73	26.41	27.08	27.76	28.44	29.11	32.50
7/8	26.25	26.98	27.71	28.44	29.17	29.90	30.63	31.35	35.00
15/16	28.13	28.91	29.69	30.47	31.25	32.03	32.81	33.59	37.50
1	30.00	30.83	31.67	32.50	33.33	34.17	35.00	35.83	40.00
1 1/16	31.88	32.76	33.65	34.53	35.42	36.30	37.19	38.07	42.50
1 1/8	33.75	34.69	35.63	36.56	37.50	38.44	39.38	40.31	45.00
1 3/16	35.63	36.61	37.60	38.59	39.58	40.57	41.56	42.55	47.50
1 1/4	37.50	38.54	39.58	40.63	41.67	42.71	43.75	44.79	50.00
1 5/16	39.38	40.47	41.56	42.66	43.75	44.84	45.94	47.03	52.50
1 3/8	41.25	42.40	43.54	44.69	45.83	46.98	48.13	49.27	55.00
1 7/16	43.13	44.32	45.52	46.72	47.92	49.11	50.31	51.51	57.50
1 1/2	45.00	46.25	47.50	48.75	50.00	51.25	52.50	53.75	60.00
1 9/16	46.88	48.18	49.48	50.78	52.08	53.39	54.69	55.99	62.50
1 5/8	48.75	50.10	51.46	52.81	54.17	55.52	56.88	58.23	65.00
1 11/16	50.63	52.03	53.44	54.84	56.25	57.66	59.06	60.47	67.50
1 3/4	52.50	53.96	55.42	56.88	58.33	59.79	61.25	62.71	70.00
1 13/16	54.38	55.89	57.40	58.91	60.42	61.93	63.44	64.95	72.50
1 7/8	56.25	57.81	59.38	60.94	62.50	64.06	65.63	67.19	75.00
1 15/16	58.13	59.74	61.35	62.97	64.58	66.20	67.81	69.43	77.50
2	60.00	61.67	63.33	65.00	66.67	68.33	70.00	71.67	80.00

WEIGHTS OF FLAT ROLLED IRON PER LINEAL FOOT.

(CONTINUED.)

Thickness in Inches.	11″	11¼″	11½″	11¾″	12″	12¼″	12½″	12¾″
1/16	2.29	2.34	2.40	2.45	2.50	2.55	2.60	2.66
1/8	4.58	4.69	4.79	4.90	5.00	5.10	5.21	5.31
3/16	6.88	7.03	7.19	7.34	7.50	7.66	7.81	7.97
1/4	9.17	9.38	9.58	9.79	10.00	10.21	10.42	10.63
5/16	11.46	11.72	11.98	12.24	12.50	12.76	13.02	13.28
3/8	13.75	14.06	14.38	14.69	15.00	15.31	15.63	15.94
7/16	16.04	16.41	16.77	17.14	17.50	17.86	18.23	18.59
1/2	18.33	18.75	19.17	19.58	20.00	20.42	20.83	21.25
9/16	20.63	21.09	21.56	22.03	22.50	22.97	23.44	23.91
5/8	22.92	23.44	23.96	24.48	25.00	25.52	26.04	26.56
11/16	25.21	25.78	26.35	26.93	27.50	28.07	28.65	29.22
3/4	27.50	28.13	28.75	29.38	30.00	30.63	31.25	31.88
13/16	29.79	30.47	31.15	31.82	32.50	33.18	33.85	34.53
7/8	32.08	32.81	33.54	34.27	35.00	35.73	36.46	37.19
15/16	34.38	35.16	35.94	36.72	37.50	38.28	39.06	39.84
1	36.67	37.50	38.33	39.17	40.00	40.83	41.67	42.50
1 1/16	38.96	39.84	40.73	41.61	42.50	43.39	44.27	45.16
1 1/8	41.25	42.19	43.13	44.06	45.00	45.94	46.88	47.81
1 3/16	43.54	44.53	45.52	46.51	47.50	48.49	49.48	50.47
1 1/4	45.83	46.88	47.92	48.96	50.00	51.04	52.08	53.13
1 5/16	48.13	49.22	50.31	51.41	52.50	53.59	54.69	55.78
1 3/8	50.42	51.56	52.71	53.85	55.00	56.15	57.29	58.44
1 7/16	52.71	53.91	55.10	56.30	57.50	58.70	59.90	61.09
1 1/2	55.00	56.25	57.50	58.75	60.00	61.25	62.50	63.75
1 9/16	57.29	58.59	59.90	61.20	62.50	63.80	65.10	66.41
1 5/8	59.58	60.94	62.29	63.65	65.00	66.35	67.71	69.06
1 11/16	61.88	63.28	64.69	66.09	67.50	68.91	70.31	71.72
1 3/4	64.17	65.63	67.08	68.54	70.00	71.46	72.92	74.38
1 13/16	66.46	67.97	69.48	70.99	72.50	74.01	75.52	77.03
1 7/8	68.75	70.31	71.88	73.44	75.00	76.56	78.13	79.69
1 15/16	71.04	72.66	74.27	75.89	77.50	79.11	80.73	82.34
2	73.33	75.00	76.67	78.33	80.00	81.67	83.33	85.00

The weights for 12″ width are repeated on each page to facilitate making the additions necessary to obtain the weights of plates wider than 12″. Thus, to find the weight of 15¾″ × ⅞″, add the weights to be found in the same line for 3¼ × ⅞ and 12 × ⅞ = 9.48 + 35.00 = 44.48 lbs.

AREAS OF FLAT ROLLED IRON,

For Thicknesses from 1/16 in. to 2 in. and Widths from 1 in. to 12¾ in.

Thickness in Inches.	1″	1¼″	1½″	1¾″	2″	2¼″	2½″	2¾″	12″
1/16	.063	.078	.094	.109	.125	.141	.156	.172	.750
1/8	.125	.156	.188	.219	.250	.281	.313	.344	1.50
3/16	.188	.234	.281	.328	.375	.422	.469	.516	2.25
1/4	.250	.313	.375	.438	.500	.563	.625	.688	3.00
5/16	.313	.391	.469	.547	.625	.703	.781	.859	3.75
3/8	.375	.469	.563	.656	.750	.844	.938	1.03	4.50
7/16	.438	.547	.656	.766	.875	.984	1.09	1.20	5.25
1/2	.500	.625	.750	.875	1.00	1.13	1.25	1.38	6.00
9/16	.563	.703	.844	.984	1.13	1.27	1.41	1.55	6.75
5/8	.625	.781	.938	1.09	1.25	1.41	1.56	1.72	7.50
11/16	.688	.859	1.03	1.20	1.38	1.55	1.72	1.89	8.25
3/4	.750	.938	1.13	1.31	1.50	1.69	1.88	2.06	9.00
13/16	.813	1.02	1.22	1.42	1.63	1.83	2.03	2.23	9.75
7/8	.875	1.09	1.31	1.53	1.75	1.97	2.19	2.41	10.50
15/16	.938	1.17	1.41	1.64	1.88	2.11	2.34	2.58	11.25
1	1.00	1.25	1.50	1.75	2.00	2.25	2.50	2.75	12.00
1 1/16	1.06	1.33	1.59	1.86	2.13	2.39	2.66	2.92	12.75
1 1/8	1.13	1.41	1.69	1.97	2.25	2.53	2.81	3.09	13.50
1 3/16	1.19	1.48	1.78	2.08	2.38	2.67	2.97	3.27	14.25
1 1/4	1.25	1.56	1.88	2.19	2.50	2.81	3.13	3.44	15.00
1 5/16	1.31	1.64	1.97	2.30	2.63	2.95	3.28	3.61	15.75
1 3/8	1.38	1.72	2.06	2.41	2.75	3.09	3.44	3.78	16.50
1 7/16	1.44	1.80	2.16	2.52	2.88	3.23	3.59	3.95	17.25
1 1/2	1.50	1.88	2.25	2.63	3.00	3.38	3.75	4.13	18.00
1 9/16	1.56	1.95	2.34	2.73	3.13	3.52	3.91	4.30	18.75
1 5/8	1.63	2.03	2.44	2.84	3.25	3.66	4.06	4.47	19.50
1 11/16	1.69	2.11	2.53	2.95	3.38	3.80	4.22	4.64	20.25
1 3/4	1.75	2.19	2.63	3.06	3.50	3.94	4.38	4.81	21.00
1 13/16	1.81	2.27	2.72	3.17	3.63	4.08	4.53	4.98	21.75
1 7/8	1.88	2.34	2.81	3.28	3.75	4.22	4.69	5.16	22.50
1 15/16	1.94	2.42	2.91	3.39	3.88	4.36	4.84	5.33	23.25
2	2.00	2.50	3.00	3.50	4.00	4.50	5.00	5.50	24.00

AREAS OF FLAT ROLLED IRON.

(CONTINUED.)

Thickness in Inches.	3″	3¼″	3½″	3¾″	4″	4¼″	4½″	4¾″	12″
1/16	.188	.203	.219	.234	.250	.266	.281	.297	.750
1/8	.375	.406	.438	.469	.500	.531	.563	.594	1.50
3/16	.563	.609	.656	.703	.750	.797	.844	.891	2.25
1/4	.750	.813	.875	.938	1.00	1.06	1.13	1.19	3.00
5/16	.938	1.02	1.09	1.17	1.25	1.33	1.41	1.48	3.75
3/8	1.13	1.22	1.31	1.41	1.50	1.59	1.69	1.78	4.50
7/16	1.31	1.42	1.53	1.64	1.75	1.86	1.97	2.08	5.25
1/2	1.50	1.63	1.75	1.88	2.00	2.13	2.25	2.38	6.00
9/16	1.69	1.83	1.97	2.11	2.25	2.39	2.53	2.67	6.75
5/8	1.88	2.03	2.19	2.34	2.50	2.66	2.81	2.97	7.50
11/16	2.06	2.23	2.41	2.58	2.75	2.92	3.09	3.27	8.25
3/4	2.25	2.44	2.63	2.81	3.00	3.19	3.38	3.56	9.00
13/16	2.44	2.64	2.84	3.05	3.25	3.45	3.66	3.86	9.75
7/8	2.63	2.84	3.06	3.28	3.50	3.72	3.94	4.16	10.50
15/16	2.81	3.05	3.28	3.52	3.75	3.98	4.22	4.45	11.25
1	3.00	3.25	3.50	3.75	4.00	4.25	4.50	4.75	12.00
1 1/16	3.19	3.45	3.72	3.98	4.25	4.52	4.78	5.05	12.75
1 1/8	3.38	3.66	3.94	4.22	4.50	4.78	5.06	5.34	13.50
1 3/16	3.56	3.86	4.16	4.45	4.75	5.05	5.34	5.64	14.25
1 1/4	3.75	4.06	4.38	4.69	5.00	5.31	5.63	5.94	15.00
1 5/16	3.94	4.27	4.59	4.92	5.25	5.58	5.91	6.23	15.75
1 3/8	4.13	4.47	4.81	5.16	5.50	5.84	6.19	6.53	16.50
1 7/16	4.31	4.67	5.03	5.39	5.75	6.11	6.47	6.83	17.25
1 1/2	4.50	4.88	5.25	5.63	6.00	6.38	6.75	7.13	18.00
1 9/16	4.69	5.08	5.47	5.86	6.25	6.64	7.03	7.42	18.75
1 5/8	4.88	5.28	5.69	6.09	6.50	6.91	7.31	7.72	19.50
1 11/16	5.06	5.48	5.91	6.33	6.75	7.17	7.59	8.02	20.25
1 3/4	5.25	5.69	6.13	6.56	7.00	7.44	7.88	8.31	21.00
1 13/16	5.44	5.89	6.34	6.80	7.25	7.70	8.16	8.61	21.75
1 7/8	5.63	6.09	6.56	7.03	7.50	7.97	8.44	8.91	22.50
1 15/16	5.81	6.30	6.78	7.27	7.75	8.23	8.72	9.20	23.25
2	6.00	6.50	7.00	7.50	8.00	8.50	9.00	9.50	24.00

AREAS OF FLAT ROLLED IRON.

(CONTINUED.)

Thickness in Inches.	5″	5¼″	5½″	5¾″	6″	6¼″	6½″	6¾″	12″
1/16	.313	.328	.344	.359	.375	.391	.406	.422	.750
1/8	.625	.656	.688	.719	.750	.781	.813	.844	1.50
3/16	.938	.984	1.03	1.08	1.13	1.17	1.22	1.27	2.25
1/4	1.25	1.31	1.38	1.44	1.50	1.56	1.63	1.69	3.00
5/16	1.56	1.64	1.72	1.80	1.88	1.95	2.03	2.11	3.75
3/8	1.88	1.97	2.06	2.16	2.25	2.34	2.44	2.53	4.50
7/16	2.19	2.30	2.41	2.52	2.63	2.73	2.84	2.95	5.25
1/2	2.50	2.63	2.75	2.88	3.00	3.13	3.25	3.38	6.00
9/16	2.81	2.95	3.09	3.23	3.38	3.52	3.66	3.80	6.75
5/8	3.13	3.28	3.44	3.59	3.75	3.91	4.06	4.22	7.50
11/16	3.44	3.61	3.78	3.95	4.13	4.30	4.47	4.64	8.25
3/4	3.75	3.94	4.13	4.31	4.50	4.69	4.88	5.06	9.00
13/16	4.06	4.27	4.47	4.67	4.88	5.08	5.28	5.48	9.75
7/8	4.38	4.59	4.81	5.03	5.25	5.47	5.69	5.91	10.50
15/16	4.69	4.92	5.16	5.39	5.63	5.86	6.09	6.33	11.25
1	5.00	5.25	5.50	5.75	6.00	6.25	6.50	6.75	12.00
1 1/16	5.31	5.58	5.84	6.11	6.38	6.64	6.91	7.17	12.75
1 1/8	5.63	5.91	6.19	6.47	6.75	7.03	7.31	7.59	13.50
1 3/16	5.94	6.23	6.53	6.83	7.13	7.42	7.72	8.02	14.25
1 1/4	6.25	6.56	6.88	7.19	7.50	7.81	8.13	8.44	15.00
1 5/16	6.56	6.89	7.22	7.55	7.88	8.20	8.53	8.86	15.75
1 3/8	6.88	7.22	7.56	7.91	8.25	8.59	8.94	9.28	16.50
1 7/16	7.19	7.55	7.91	8.27	8.63	8.98	9.34	9.70	17.25
1 1/2	7.50	7.88	8.25	8.63	9.00	9.38	9.75	10.13	18.00
1 9/16	7.81	8.20	8.59	8.98	9.38	9.77	10.16	10.55	18.75
1 5/8	8.13	8.53	8.94	9.34	9.75	10.16	10.56	10.97	19.50
1 11/16	8.44	8.86	9.28	9.70	10.13	10.55	10.97	11.39	20.25
1 3/4	8.75	9.19	9.63	10.06	10.50	10.94	11.38	11.81	21.00
1 13/16	9.06	9.52	9.97	10.42	10.88	11.33	11.78	12.23	21.75
1 7/8	9.38	9.84	10.31	10.78	11.25	11.72	12.19	12.66	22.50
1 15/16	9.69	10.17	10.66	11.14	11.63	12.11	12.59	13.08	23.25
2	10.00	10.50	11.00	11.50	12.00	12.50	13.00	13.50	24.00

AREAS OF FLAT ROLLED IRON.

(CONTINUED.)

Thickness in Inches.	7″	7¼″	7½″	7¾″	8″	8¼″	8½″	8¾″	12″
1/16	.438	.453	.469	.484	.500	.516	.531	.547	.750
1/8	.875	.906	.938	.969	1.00	1.03	1.06	1.09	1.50
3/16	1.31	1.36	1.41	1.45	1.50	1.55	1.59	1.64	2.25
1/4	1.75	1.81	1.88	1.94	2.00	2.06	2.13	2.19	3.00
5/16	2.19	2.27	2.34	2.42	2.50	2.58	2.66	2.73	3.75
3/8	2.63	2.72	2.81	2.91	3.00	3.09	3.19	3.28	4.50
7/16	3.06	3.17	3.28	3.39	3.50	3.61	3.72	3.83	5.25
1/2	3.50	3.63	3.75	3.88	4.00	4.13	4.25	4.38	6.00
9/16	3.94	4.08	4.22	4.36	4.50	4.64	4.78	4.92	6.75
5/8	4.38	4.53	4.69	4.84	5.00	5.16	5.31	5.47	7.50
11/16	4.81	4.98	5.16	5.33	5.50	5.67	5.84	6.02	8.25
3/4	5.25	5.44	5.63	5.81	6.00	6.19	6.38	6.56	9.00
13/16	5.69	5.89	6.09	6.30	6.50	6.70	6.91	7.11	9.75
7/8	6.13	6.34	6.56	6.78	7.00	7.22	7.44	7.66	10.50
15/16	6.56	6.80	7.03	7.27	7.50	7.73	7.97	8.20	11.25
1	7.00	7.25	7.50	7.75	8.00	8.25	8.50	8.75	12.00
1 1/16	7.44	7.70	7.97	8.23	8.50	8.77	9.03	9.30	12.75
1 1/8	7.88	8.16	8.44	8.72	9.00	9.28	9.56	9.84	13.50
1 3/16	8.31	8.61	8.91	9.20	9.50	9.80	10.09	10.39	14.25
1 1/4	8.75	9.06	9.38	9.69	10.00	10.31	10.63	10.94	15.00
1 5/16	9.19	9.52	9.84	10.17	10.50	10.83	11.16	11.48	15.75
1 3/8	9.63	9.97	10.31	10.66	11.00	11.34	11.69	12.03	16.50
1 7/16	10.06	10.42	10.78	11.14	11.50	11.86	12.22	12.58	17.25
1 1/2	10.50	10.88	11.25	11.63	12.00	12.38	12.75	13.13	18.00
1 9/16	10.94	11.33	11.72	12.11	12.50	12.89	13.28	13.67	18.75
1 5/8	11.38	11.78	12.19	12.59	13.00	13.41	13.81	14.22	19.50
1 11/16	11.81	12.23	12.66	13.08	13.50	13.92	14.34	14.77	20.25
1 3/4	12.25	12.69	13.13	13.56	14.00	14.44	14.88	15.31	21.00
1 13/16	12.69	13.14	13.59	14.05	14.50	14.95	15.41	15.86	21.75
1 7/8	13.13	13.59	14.06	14.53	15.00	15.47	15.94	16.41	22.50
1 15/16	13.56	14.05	14.53	15.02	15.50	15.98	16.47	16.95	23.25
2	14.00	14.50	15.00	15.50	16.00	16.50	17.00	17.50	24.00

AREAS OF FLAT ROLLED IRON.

(CONTINUED.)

Thickness in Inches.	9″	9¼″	9½″	9¾″	10″	10¼″	10½″	10¾″	12″
1/16	.563	.578	.594	.609	.625	.641	.656	.672	.750
1/8	1.13	1.16	1.19	1.22	1.25	1.28	1.31	1.34	1.50
3/16	1.69	1.73	1.78	1.83	1.88	1.92	1.97	2.02	2.25
1/4	2.25	2.31	2.38	2.44	2.50	2.56	2.63	2.69	3.00
5/16	2.81	2.89	2.97	3.05	3.13	3.20	3.28	3.36	3.75
3/8	3.38	3.47	3.56	3.66	3.75	3.84	3.94	4.03	4.50
7/16	3.94	4.05	4.16	4.27	4.38	4.48	4.59	4.70	5.25
1/2	4.50	4.63	4.75	4.88	5.00	5.13	5.25	5.38	6.00
9/16	5.06	5.20	5.34	5.48	5.63	5.77	5.91	6.05	6.75
5/8	5.63	5.78	5.94	6.09	6.25	6.41	6.56	6.72	7.50
11/16	6.19	6.36	6.53	6.70	6.88	7.05	7.22	7.39	8.25
3/4	6.75	6.94	7.13	7.31	7.50	7.69	7.88	8.06	9.00
13/16	7.31	7.52	7.72	7.92	8.13	8.33	8.53	8.73	9.75
7/8	7.88	8.09	8.31	8.53	8.75	8.97	9.19	9.41	10.50
15/16	8.44	8.67	8.91	9.14	9.38	9.61	9.84	10.08	11.25
1	9.00	9.25	9.50	9.75	10.00	10.25	10.50	10.75	12.00
1 1/16	9.56	9.83	10.09	10.36	10.63	10.89	11.16	11.42	12.75
1 1/8	10.13	10.41	10.69	10.97	11.25	11.53	11.81	12.09	13.50
1 3/16	10.69	10.98	11.28	11.58	11.88	12.17	12.47	12.77	14.25
1 1/4	11.25	11.56	11.88	12.19	12.50	12.81	13.13	13.44	15.00
1 5/16	11.81	12.14	12.47	12.80	13.13	13.45	13.78	14.11	15.75
1 3/8	12.38	12.72	13.06	13.41	13.75	14.09	14.44	14.78	16.50
1 7/16	12.94	13.30	13.66	14.02	14.38	14.73	15.09	15.45	17.25
1 1/2	13.50	13.88	14.25	14.63	15.00	15.38	15.75	16.13	18.00
1 9/16	14.06	14.45	14.84	15.23	15.63	16.02	16.41	16.80	18.75
1 5/8	14.63	15.03	15.44	15.84	16.25	16.66	17.06	17.47	19.50
1 11/16	15.19	15.61	16.03	16.45	16.88	17.30	17.72	18.14	20.25
1 3/4	15.75	16.19	16.63	17.06	17.50	17.94	18.38	18.81	21.00
1 13/16	16.31	16.77	17.22	17.67	18.13	18.58	19.03	19.48	21.75
1 7/8	16.88	17.34	17.81	18.28	18.75	19.22	19.69	20.16	22.50
1 15/16	17.44	17.92	18.41	18.89	19.38	19.86	20.34	20.83	23.25
2	18.00	18.50	19.00	19.50	20.00	20.50	21.00	21.50	24.00

SQUARE AND ROUND BARS.

(CONTINUED.)

Thickness or Diameter in Inches.	Weight of ☐ Bar One Foot long.	Weight of ○ Bar One Foot long.	Area of ☐ Bar in sq. inches.	Area of ○ Bar in sq. inches.	Circumference of ○ Bar in inches.
2	13.33	10.47	4.0000	3.1416	6.2832
1/16	14.18	11.14	4.2539	3.3410	6.4795
1/8	15.05	11.82	4.5156	3.5466	6.6759
3/16	15.95	12.53	4.7852	3.7583	6.8722
1/4	16.88	13.25	5.0625	3.9761	7.0686
5/16	17.83	14.00	5.3477	4.2000	7.2649
3/8	18.80	14.77	5.6406	4.4301	7.4613
7/16	19.80	15.55	5.9414	4.6664	7.6576
1/2	20.83	16.36	6.2500	4.9087	7.8540
9/16	21.89	17.19	6.5664	5.1572	8.0503
5/8	22.97	18.04	6.8906	5.4119	8.2467
11/16	24.08	18.91	7.2227	5.6727	8.4430
3/4	25.21	19.80	7.5625	5.9396	8.6394
13/16	26.37	20.71	7.9102	6.2126	8.8357
7/8	27.55	21.64	8.2656	6.4918	9.0321
15/16	28.76	22.59	8.6289	6.7771	9.2284
3	30.00	23.56	9.0000	7.0686	9.4248
1/16	31.26	24.55	9.3789	7.3662	9.6211
1/8	32.55	25.57	9.7656	7.6699	9.8175
3/16	33.87	26.60	10.160	7.9798	10.014
1/4	35.21	27.65	10.563	8.2958	10.210
5/16	36.58	28.73	10.973	8.6179	10.407
3/8	37.97	29.82	11.391	8.9462	10.603
7/16	39.39	30.94	11.816	9.2806	10.799
1/2	40.83	32.07	12.250	9.6211	10.996
9/16	42.30	33.23	12.691	9.9678	11.192
5/8	43.80	34.40	13.141	10.321	11.388
11/16	45.33	35.60	13.598	10.680	11.585
3/4	46.88	36.82	14.063	11.045	11.781
13/16	48.45	38.05	14.535	11.416	11.977
7/8	50.05	39.31	15.016	11.793	12.174
15/16	51.68	40.59	15.504	12.177	12.370

SQUARE AND ROUND BARS.

(CONTINUED.)

Thickness or Diameter in Inches.	Weight of ☐ Bar One Foot long.	Weight of ○ Bar One Foot long.	Area of ☐ Bar in sq. inches.	Area of ○ Bar in sq. inches.	Circumference of ○ Bar in inches.
4	53.33	41.89	16.000	12.566	12.566
1/16	55.01	43.21	16.504	12.962	12.763
1/8	56.72	44.55	17.016	13.364	12.959
3/16	58.45	45.91	17.535	13.772	13.155
1/4	60.21	47.29	18.063	14.186	13.352
5/16	61.99	48.69	18.598	14.607	13.548
3/8	63.80	50.11	19.141	15.033	13.744
7/16	65.64	51.55	19.691	15.466	13.941
1/2	67.50	53.01	20.250	15.904	14.137
9/16	69.39	54.50	20.816	16.349	14.334
5/8	71.30	56.00	21.391	16.800	14.530
11/16	73.24	57.52	21.973	17.257	14.726
3/4	75.21	59.07	22.563	17.721	14.923
13/16	77.20	60.63	23.160	18.190	15.119
7/8	79.22	62.22	23.766	18.665	15.315
15/16	81.26	63.82	24.379	19.147	15.512
5	83.33	65.45	25.000	19.635	15.708
1/16	85.43	67.10	25.629	20.129	15.904
1/8	87.55	68.76	26.266	20.629	16.101
3/16	89.70	70.45	26.910	21.135	16.297
1/4	91.88	72.16	27.563	21.648	16.493
5/16	94.08	73.89	28.223	22.166	16.690
3/8	96.30	75.64	28.891	22.691	16.886
7/16	98.55	77.40	29.566	23.221	17.082
1/2	100.8	79.19	30.250	23.758	17.279
9/16	103.1	81.00	30.941	24.301	17.475
5/8	105.5	82.83	31.641	24.850	17.671
11/16	107.8	84.69	32.348	25.406	17.868
3/4	110.2	86.56	33.063	25.967	18.064
13/16	112.6	88.45	33.785	26.535	18.261
7/8	115.1	90.36	34.516	27.109	18.457
15/16	117.5	92.29	35.254	27.688	18.653

SQUARE AND ROUND BARS.

(CONTINUED.)

Thickness or Diameter in Inches.	Weight of ☐ Bar One Foot long.	Weight of ○ Bar One Foot long.	Area of ☐ Bar in sq. inches.	Area of ○ Bar in sq. inches.	Circumference of ○ Bar in inches.
6	120.0	94.25	36.000	28.274	18.850
1/16	122.5	96.22	36.754	28.866	19.046
1/8	125.1	98.22	37.516	29.465	19.242
3/16	127.6	100.2	38.285	30.069	19.439
1/4	130.2	102.3	39.063	30.680	19.635
5/16	132.8	104.3	39.848	31.296	19.831
3/8	135.5	106.4	40.641	31.919	20.028
7/16	138.1	108.5	41.441	32.548	20.224
1/2	140.8	110.6	42.250	33.183	20.420
9/16	143.6	112.7	43.066	33.824	20.617
5/8	146.3	114.9	43.891	34.472	20.813
11/16	149.1	117.1	44.723	35.125	21.009
3/4	151.9	119.3	45.563	35.785	21.206
13/16	154.7	121.5	46.410	36.450	21.402
7/8	157.6	123.7	47.266	37.122	21.598
15/16	160.4	126.0	48.129	37.800	21.795
7	163.3	128.3	49.000	38.485	21.991
1/16	166.3	130.6	49.879	39.175	22.187
1/8	169.2	132.9	50.766	39.871	22.384
3/16	172.2	135.2	51.660	40.574	22.580
1/4	175.2	137.6	52.563	41.282	22.777
5/16	178.2	140.0	53.473	41.997	22.973
3/8	181.3	142.4	54.391	42.718	23.169
7/16	184.4	144.8	55.316	43.445	23.366
1/2	187.5	147.3	56.250	44.179	23.562
9/16	190.6	149.7	57.191	44.918	23.758
5/8	193.8	152.2	58.141	45.664	23.955
11/16	197.0	154.7	59.098	46.415	24.151
3/4	200.2	157.2	60.063	47.173	24.347
13/16	203.5	159.8	61.035	47.937	24.544
7/8	206.7	162.4	62.016	48.707	24.740
15/16	210.0	164.9	63.004	49.483	24.936

SQUARE AND ROUND BARS.

(CONTINUED.)

Thickness or Diameter in Inches.	Weight of ☐ Bar One Foot long.	Weight of ◯ Bar One Foot long.	Area of ☐ Bar in sq. inches.	Area of ◯ Bar in sq. inches.	Circumference of ◯ Bar in inches.
8	213.3	167.6	64.000	50.265	25.133
1/16	216.7	170.2	65.004	51.054	25.329
1/8	220.1	172.8	66.016	51.849	25.525
3/16	223.5	175.5	67.035	52.649	25.722
1/4	226.9	178.2	68.063	53.456	25.918
5/16	230.3	180.9	69.098	54.269	26.114
3/8	233.8	183.6	70.141	55.088	26.311
7/16	237.3	186.4	71.191	55.914	26.507
1/2	240.8	189.2	72.250	56.745	26.704
9/16	244.4	191.9	73.316	57.583	26.900
5/8	248.0	194.8	74.391	58.426	27.096
11/16	251.6	197.6	75.473	59.276	27.293
3/4	255.2	200.4	76.563	60.132	27.489
13/16	258.9	203.3	77.660	60.994	27.685
7/8	262.6	206.2	78.766	61.862	27.882
15/16	266.3	209.1	79.879	62.737	28.078
9	270.0	212.1	81.000	63.617	28.274
1/16	273.8	215.0	82.129	64.504	28.471
1/8	277.6	218.0	83.266	65.397	28.667
3/16	281.4	221.0	84.410	66.296	28.863
1/4	285.2	224.0	85.563	67.201	29.060
5/16	289.1	227.0	86.723	68.112	29.256
3/8	293.0	230.1	87.891	69.029	29.452
7/16	296.9	233.2	89.066	69.953	29.649
1/2	300.8	236.3	90.250	70.882	29.845
9/16	304.8	239.4	91.441	71.818	30.041
5/8	308.8	242.5	92.641	72.760	30.238
11/16	312.8	245.7	93.848	73.708	30.434
3/4	316.9	248.9	95.063	74.662	30.631
13/16	321.0	252.1	96.285	75.622	30.827
7/8	325.1	255.3	97.516	76.589	31.023
15/16	329.2	258.5	98.754	77.561	31.220

SQUARE AND ROUND BARS.
(CONTINUED.)

Thickness or Diameter in Inches.	Weight of □ Bar One Foot long.	Weight of ○ Bar One Foot long.	Area of □ Bar in sq. inches.	Area of ○ Bar in sq. inches.	Circumference of ○ Bar in inches.
10	333.3	261.8	100.00	78.540	31.416
1/16	337.5	265.1	101.25	79.525	31.612
1/8	341.7	268.4	102.52	80.516	31.809
3/16	346.0	271.7	103.79	81.513	32.005
1/4	350.2	275.1	105.06	82.516	32.201
5/16	354.5	278.4	106.35	83.525	32.398
3/8	358.8	281.8	107.64	84.541	32.594
7/16	363.1	285.2	108.94	85.562	32.790
1/2	367.5	288.6	110.25	86.590	32.987
9/16	371.9	292.1	111.57	87.624	33.183
5/8	376.3	295.5	112.89	88.664	33.379
11/16	380.7	299.0	114.22	89.710	33.576
3/4	385.2	302.5	115.56	90.763	33.772
13/16	389.7	306.1	116.91	91.821	33.968
7/8	394.2	309.6	118.27	92.886	34.165
15/16	398.8	313.2	119.63	93.956	34.361
11	403.3	316.8	121.00	95.033	34.558
1/16	407.9	320.4	122.38	96.116	34.754
1/8	412.6	324.0	123.77	97.205	34.950
3/16	417.2	327.7	125.16	98.301	35.147
1/4	421.9	331.3	126.56	99.402	35.343
5/16	426.6	335.0	127.97	100.51	35.539
3/8	431.3	338.7	129.39	101.62	35.736
7/16	436.1	342.5	130.82	102.74	35.932
1/2	440.8	346.2	132.25	103.87	36.128
9/16	445.6	350.0	133.69	105.00	36.325
5/8	450.5	353.8	135.14	106.14	36.521
11/16	455.3	357.6	136.60	107.28	36.717
3/4	460.2	361.4	138.06	108.43	36.914
13/16	465.1	365.3	139.54	109.59	37.110
7/8	470.1	369.2	141.02	110.75	37.306
15/16	475.0	373.1	142.50	111.92	37.503

WEIGHT OF SHEETS OF WROUGHT IRON, STEEL, COPPER AND BRASS. (From Haswell.)

Weights per Square Foot. Thickness by Birmingham Gauge.

No. of Gauge.	Thickness in inches.	Iron.	Steel.	Copper.	Brass.
0000	.454	18.22	18.46	20.57	19.43
000	.425	17.05	17.28	19.25	18.19
00	.38	15.25	15.45	17.21	16.26
0	.34	13.64	13.82	15.40	14.55
1	.3	12.04	12.20	13.59	12.84
2	.284	11.40	11.55	12.87	12.16
3	.259	10.39	10.53	11.73	11.09
4	.238	9.55	9.68	10.78	10.19
5	.22	8.83	8.95	9.97	9.42
6	.203	8.15	8.25	9.20	8.69
7	.18	7.22	7.32	8.15	7.70
8	.165	6.62	6.71	7.47	7.06
9	.148	5.94	6.02	6.70	6.33
10	.134	5.38	5.45	6.07	5.74
11	.12	4.82	4.88	5.44	5.14
12	.109	4.37	4.43	4.94	4.67
13	.095	3.81	3.86	4.30	4.07
14	.083	3.33	3.37	3.76	3.55
15	.072	2.89	2.93	3.26	3.08
16	.065	2.61	2.64	2.94	2.78
17	.058	2.33	2.36	2.63	2.48
18	.049	1.97	1.99	2.22	2.10
19	.042	1.69	1.71	1.90	1.80
20	.035	1.40	1.42	1.59	1.50
21	.032	1.28	1.30	1.45	1.37
22	.028	1.12	1.14	1.27	1.20
23	.025	1.00	1.02	1.13	1.07
24	.022	.883	.895	1.00	.942
25	.02	.803	.813	.906	.856
26	.018	.722	.732	.815	.770
27	.016	.642	.651	.725	.685
28	.014	.562	.569	.634	.599
29	.013	.522	.529	.589	.556
30	.012	.482	.488	.544	.514
31	.01	.401	.407	.453	.428
32	.009	.361	.366	.408	.385
33	.008	.321	.325	.362	.342
34	.007	.281	.285	.317	.300
35	.005	.201	.203	.227	.214
Specific Gravity,		7.704	7.806	8.698	8.218
Weight Cubic Foot,		481.25	487.75	543.6	513.6
" " Inch,		.2787	.2823	.3146	.2972

WEIGHT OF SHEETS OF WROUGHT IRON, STEEL, COPPER AND BRASS. (From Haswell.)

Weights per Sq. Foot. Thickness by American (Browne & Sharpe's) Gauge.

No. of Gauge.	Thickness in inches.	Iron.	Steel.	Copper.	Brass.
0000	.46	18.46	18.70	20.84	19.69
000	.4096	16.44	16.66	18.56	17.53
00	.3648	14.64	14.83	16.53	15.61
0	.3249	13.04	13.21	14.72	13.90
1	.2893	11.61	11.76	13.11	12.38
2	.2576	10.34	10.48	11.67	11.03
3	.2294	9.21	9.33	10.39	9.82
4	.2043	8.20	8.31	9.26	8.74
5	.1819	7.30	7.40	8.24	7.79
6	.1620	6.50	6.59	7.34	6.93
7	.1443	5.79	5.87	6.54	6.18
8	.1285	5.16	5.22	5.82	5.50
9	.1144	4.59	4.65	5.18	4.90
10	.1019	4.09	4.14	4.62	4.36
11	.0907	3.64	3.69	4.11	3.88
12	.0808	3.24	3.29	3.66	3.46
13	.0720	2.89	2.93	3.26	3.08
14	.0641	2.57	2.61	2.90	2.74
15	.0571	2.29	2.32	2.59	2.44
16	.0508	2.04	2.07	2.30	2.18
17	.0453	1.82	1.84	2.05	1.94
18	.0403	1.62	1.64	1.83	1.73
19	.0359	1.44	1.46	1.63	1.54
20	.0320	1.28	1.30	1.45	1.37
21	.0285	1.14	1.16	1.29	1.22
22	.0253	1.02	1.03	1.15	1.08
23	.0226	.906	.918	1.02	.966
24	.0201	.807	.817	.911	.860
25	.0179	.718	.728	.811	.766
26	.0159	.640	.648	.722	.682
27	.0142	.570	.577	.643	.608
28	.0126	.507	.514	.573	.541
29	.0113	.452	.458	.510	.482
30	.0100	.402	.408	.454	.429
31	.0089	.358	.363	.404	.382
32	.0080	.319	.323	.360	.340
33	.0071	.284	.288	.321	.303
34	.0063	.253	.256	.286	.270
35	.0056	.225	.228	.254	.240

As there are many gauges in use differing from each other, and even the thicknesses of a certain specified gauge, as the Birmingham, are not assumed the same by all manufacturers, orders for sheets and wire should always state the weight per square foot, or the thickness in thousandths of an inch.

AREAS and CIRCUMFERENCES OF CIRCLES.

For Diameters from $\frac{1}{10}$ to 100, advancing by Tenths.

Diam.	Area.	Circum.	Diam.	Area.	Circum.
0.0			4.0	12.5664	12.5664
.1	.007854	.31416	.1	13.2025	12.8805
.2	.031416	.62832	.2	13.8544	13.1947
.3	.070686	.94248	.3	14.5220	13.5088
.4	.12566	1.2566	.4	15.2053	13.8230
.5	.19635	1.5708	.5	15.9043	14.1372
.6	.28274	1.8850	.6	16.6190	14.4513
.7	.38485	2.1991	.7	17.3494	14.7655
.8	.50266	2.5133	.8	18.0956	15.0796
.9	.63617	2.8274	.9	18.8574	15.3938
1.0	.7854	3.1416	5.0	19.6350	15.7080
.1	.9503	3.4558	.1	20.4282	16.0221
.2	1.1310	3.7699	.2	21.2372	16.3363
.3	1.3273	4.0841	.3	22.0618	16.6504
.4	1.5394	4.3982	.4	22.9022	16.9646
.5	1.7671	4.7124	.5	23.7583	17.2788
.6	2.0106	5.0265	.6	24.6301	17.5929
.7	2.2698	5.3407	.7	25.5176	17.9071
.8	2.5447	5.6549	.8	26.4208	18.2212
.9	2.8353	5.9690	.9	27.3397	18.5354
2.0	3.1416	6.2832	6.0	28.2743	18.8496
.1	3.4636	6.5973	.1	29.2247	19.1637
.2	3.8013	6.9115	.2	30.1907	19.4779
.3	4.1548	7.2257	.3	31.1725	19.7920
.4	4.5239	7.5398	.4	32.1699	20.1062
.5	4.9087	7.8540	.5	33.1831	20.4204
.6	5.3093	8.1681	.6	34.2119	20.7345
.7	5.7256	8.4823	.7	35.2565	21.0487
.8	6.1575	8.7965	.8	36.3168	21.3628
.9	6.6052	9.1106	.9	37.3928	21.6770
3.0	7.0686	9.4248	7.0	38.4845	21.9911
.1	7.5477	9.7389	.1	39.5919	22.3053
.2	8.0425	10.0531	.2	40.7150	22.6195
.3	8.5530	10.3673	.3	41.8539	22.9336
.4	9.0792	10.6814	.4	43.0084	23.2478
.5	9.6211	10.9956	.5	44.1786	23.5619
.6	10.1788	11.3097	.6	45.3646	23.8761
.7	10.7521	11.6239	.7	46.5663	24.1903
.8	11.3411	11.9381	.8	47.7836	24.5044
.9	11.9459	12.2522	.9	49.0167	24.8186

AREAS and CIRCUMFERENCES OF CIRCLES.
(CONTINUED.)

Diam.	Area.	Circum.	Diam.	Area.	Circum.
8.0	50.2655	25.1327	12.0	113.0973	37.6991
.1	51.5300	25.4469	.1	114.9901	38.0133
.2	52.8102	25.7611	.2	116.8987	38.3274
.3	54.1061	26.0752	.3	118.8229	38.6416
.4	55.4177	26.3894	.4	120.7628	38.9557
.5	56.7450	26.7035	.5	122.7185	39.2699
.6	58.0880	27.0177	.6	124.6898	39.5841
.7	59.4468	27.3319	.7	126.6769	39.8982
.8	60.8212	27.6460	.8	128.6796	40.2124
.9	62.2114	27.9602	.9	130.6981	40.5265
9.0	63.6173	28.2743	13.0	132.7323	40.8407
.1	65.0388	28.5885	.1	134.7822	41.1549
.2	66.4761	28.9027	.2	136.8478	41.4690
.3	67.9291	29.2168	.3	138.9291	41.7832
.4	69.3978	29.5310	.4	141.0261	42.0973
.5	70.8822	29.8451	.5	143.1388	42.4115
.6	72.3823	30.1593	.6	145.2672	42.7257
.7	73.8981	30.4734	.7	147.4114	43.0398
.8	75.4296	30.7876	.8	149.5712	43.3540
.9	76.9769	31.1018	.9	151.7468	43.6681
10.0	78.5398	31.4159	14.0	153.9380	43.9823
.1	80.1185	31.7301	.1	156.1450	44.2965
.2	81.7128	32.0442	.2	158.3677	44.6106
.3	83.3229	32.3584	.3	160.6061	44.9248
.4	84.9487	32.6726	.4	162.8602	45.2389
.5	86.5901	32.9867	.5	165.1300	45.5531
.6	88.2473	33.3009	.6	167.4155	45.8673
.7	89.9202	33.6150	.7	169.7167	46.1814
.8	91.6088	33.9292	.8	172.0336	46.4956
.9	93.3132	34.2434	.9	174.3662	46.8097
11.0	95.0332	34.5575	15.0	176.7146	47.1239
.1	96.7689	34.8717	.1	179.0786	47.4380
.2	98.5203	35.1858	.2	181.4584	47.7522
.3	100.2875	35.5000	.3	183.8539	48.0664
.4	102.0703	35.8142	.4	186.2650	48.3805
.5	103.8689	36.1283	.5	188.6919	48.6947
.6	105.6832	36.4425	.6	191.1345	49.0088
.7	107.5132	36.7566	.7	193.5928	49.3230
.8	109.3588	37.0708	.8	196.0668	49.6372
.9	111.2202	37.3850	.9	198.5565	49.9513

AREAS and CIRCUMFERENCES OF CIRCLES.

(CONTINUED.)

Diam.	Area.	Circum.	Diam.	Area.	Circum.
16.0	201.0619	50.2655	20.0	314.1593	62.8319
.1	203.5831	50.5796	.1	317.3087	63.1460
.2	206.1199	50.8938	.2	320.4739	63.4602
.3	208.6724	51.2080	.3	323.6547	63.7743
.4	211.2407	51.5221	.4	326.8513	64.0885
.5	213.8246	51.8363	.5	330.0636	64.4026
.6	216.4243	52.1504	.6	333.2916	64.7168
.7	219.0397	52.4646	.7	336.5353	65.0310
.8	221.6708	52.7788	.8	339.7947	65.3451
.9	224.3176	53.0929	.9	343.0698	65.6593
17.0	226.9801	53.4071	21.0	346.3606	65.9734
.1	229.6583	53.7212	.1	349.6671	66.2876
.2	232.3522	54.0354	.2	352.9894	66.6018
.3	235.0618	54.3496	.3	356.3273	66.9159
.4	237.7871	54.6637	.4	359.6809	67.2301
.5	240.5282	54.9779	.5	363.0503	67.5442
.6	243.2849	55.2920	.6	366.4354	67.8584
.7	246.0574	55.6062	.7	369.8361	68.1726
.8	248.8456	55.9203	.8	373.2526	68.4867
.9	251.6494	56.2345	.9	376.6848	68.8009
18.0	254.4690	56.5486	22.0	380.1327	69.1150
.1	257.3043	56.8628	.1	383.5963	69.4292
.2	260.1553	57.1770	.2	387.0756	69.7434
.3	263.0220	57.4911	.3	390.5707	70.0575
.4	265.9044	57.8053	.4	394.0814	70.3717
.5	268.8025	58.1195	.5	397.6078	70.6858
.6	271.7164	58.4336	.6	401.1500	71.0000
.7	274.6459	58.7478	.7	404.7078	71.3142
.8	277.5911	59.0619	.8	408.2814	71.6283
.9	280.5521	59.3761	.9	411.8707	71.9425
19.0	283.5287	59.6903	23.0	415.4756	72.2566
.1	286.5211	60.0044	.1	419.0963	72.5708
.2	289.5292	60.3186	.2	422.7327	72.8849
.3	292.5530	60.6327	.3	426.3848	73.1991
.4	295.5925	60.9469	.4	430.0526	73.5133
.5	298.6477	61.2611	.5	433.7361	73.8274
.6	301.7186	61.5752	.6	437.4354	74.1416
.7	304.8052	61.8894	.7	441.1503	74.4557
.8	307.9075	62.2035	.8	444.8809	74.7699
.9	311.0255	62.5177	.9	448.6273	75.0841

AREAS and CIRCUMFERENCES OF CIRCLES.

(CONTINUED.)

Diam.	Area.	Circum.	Diam.	Area.	Circum.
24.0	452.3893	75.3982	28.0	615.7522	87.9646
.1	456.1671	75.7124	.1	620.1582	88.2788
.2	459.9606	76.0265	.2	624.5800	88.5929
.3	463.7698	76.3407	.3	629.0175	88.9071
.4	467.5947	76.6549	.4	633.4707	89.2212
.5	471.4352	76.9690	.5	637.9397	89.5354
.6	475.2916	77.2832	.6	642.4243	89.8495
.7	479.1636	77.5973	.7	646.9246	90.1637
.8	483.0513	77.9115	.8	651.4407	90.4779
.9	486.9547	78.2257	9	655.9724	90.7920
25.0	490.8739	78.5398	29.0	660.5199	91.1062
.1	494.8087	78.8540	.1	665.0830	91.4203
.2	498.7592	79.1681	.2	669.6619	91.7345
.3	502.7255	79.4823	.3	674.2565	92.0487
.4	506.7075	79.7965	.4	678.8668	92.3628
.5	510.7052	80.1106	.5	683.4928	92.6770
.6	514.7185	80.4248	.6	688.1345	92.9911
.7	518.7476	80.7389	.7	692.7919	93.3053
.8	522.7924	81.0531	.8	697.4650	93.6195
.9	526.8529	81.3672	.9	702.1538	93.9336
26.0	530.9292	81.6814	30.0	706.8583	94.2478
.1	535.0211	81.9956	.1	711.5786	94.5619
.2	539.1287	82.3097	.2	716.3145	94.8761
.3	543.2521	82.6239	.3	721.0662	95.1903
.4	547.3911	82.9380	.4	725.8336	95.5044
.5	551.5459	83.2522	.5	730.6167	95.8186
.6	555.7163	83.5664	.6	735.4154	96.1327
.7	559.9025	83.8805	.7	740.2299	96.4469
.8	564.1044	84.1947	.8	745.0601	96.7611
.9	568.3220	84.5088	.9	749.9060	97.0752
27.0	572.5553	84.8230	31.0	754.7676	97.3894
.1	576.8043	85.1372	.1	759.6450	97.7035
.2	581.0690	85.4513	.2	764.5380	98.0177
.3	585.3494	85.7655	.3	769.4467	98.3319
.4	589.6455	86.0796	.4	774.3712	98.6460
.5	593.9574	86.3938	.5	779.3113	98.9602
.6	598.2849	86.7080	.6	784.2672	99.2743
.7	602.6282	87.0221	.7	789.2388	99.5885
.8	606.9871	87.3363	.8	794.2260	99.9026
.9	611.3618	87.6504	.9	799.2290	100.2168

AREAS and CIRCUMFERENCES OF CIRCLES.

(CONTINUED.)

Diam.	Area.	Circum.	Diam.	Area.	Circum.
32.0	804.2477	100.5310	36.0	1017.8760	113.0973
.1	809.2821	100.8451	.1	1023.5387	113.4115
.2	814.3322	101.1593	.2	1029.2172	113.7257
.3	819.3980	101.4734	.3	1034.9113	114.0398
.4	824.4796	101.7876	.4	1040.6212	114.3540
.5	829.5768	102.1018	.5	1046.3467	114.6681
.6	834.6898	102.4159	.6	1052.0880	114.9823
.7	839.8185	102.7301	.7	1057.8449	115.2965
.8	844.9628	103.0442	.8	1063.6176	115.6106
.9	850.1229	103.3584	.9	1069.4060	115.9248
33.0	855.2986	103.6726	37.0	1075.2101	116.2389
.1	860.4902	103.9867	.1	1081.0299	116.5531
.2	865.6973	104.3009	.2	1086.8654	116.8672
.3	870.9202	104.6150	.3	1092.7166	117.1814
.4	876.1588	104.9292	.4	1098.5835	117.4956
.5	881.4131	105.2434	.5	1104.4662	117.8097
.6	886.6831	105.5575	.6	1110.3645	118.1239
.7	891.9688	105.8717	.7	1116.2786	118.4380
.8	897.2703	106.1858	.8	1122.2083	118.7522
.9	902.5874	106.5000	.9	1128.1538	119.0664
34.0	907.9203	106.8142	38.0	1134.1149	119.3805
.1	913.2688	107.1283	.1	1140.0918	119.6947
.2	918.6331	107.4425	.2	1146.0844	120.0088
.3	924.0131	107.7566	.3	1152.0927	120.3230
.4	929.4088	108.0708	.4	1158.1167	120.6372
.5	934.8202	108.3849	.5	1164.1564	120.9513
.6	940.2473	108.6991	.6	1170.2118	121.2655
.7	945.6901	109.0133	.7	1176.2830	121.5796
.8	951.1486	109.3274	.8	1182.3698	121.8938
.9	956.6228	109.6416	.9	1188.4724	122.2080
35.0	962.1128	109.9557	39.0	1194.5906	122.5221
.1	967.6184	110.2699	.1	1200.7246	122.8363
.2	973.1397	110.5841	.2	1206.8742	123.1504
.3	978.6768	110.8982	.3	1213.0396	123.4646
.4	984.2296	111.2124	.4	1219.2207	123.7788
.5	989.7980	111.5265	.5	1225.4175	124.0929
.6	995.3822	111.8407	.6	1231.6300	124.4071
.7	1000.9821	112.1549	.7	1237.8582	124.7212
.8	1006.5977	112.4690	.8	1244.1021	125.0354
.9	1012.2290	112.7832	.9	1250.3617	125.3495

AREAS and CIRCUMFERENCES OF CIRCLES.
(CONTINUED.)

Diam.	Area.	Circum.	Diam.	Area.	Circum.
40.0	1256.6371	125.6637	44.0	1520.5308	138.2301
.1	1262.9281	125.9779	.1	1527.4502	138.5442
.2	1269.2348	126.2920	.2	1534.3853	138.8584
.3	1275.5573	126.6062	.3	1541.3360	139.1726
.4	1281.8955	126.9203	.4	1548.3025	139.4867
.5	1288.2493	127.2345	.5	1555.2847	139.8009
.6	1294.6189	127.5487	.6	1562.2826	140.1153
.7	1301.0042	127.8628	.7	1569.2962	140.4292
.8	1307.4052	128.1770	.8	1576.3255	140.7434
.9	1313.8219	128.4911	.9	1583.3706	141.0575
41.0	1320.2543	128.8053	45.0	1590.4313	141.3717
.1	1326.7024	129.1195	.1	1597.5077	141.6858
.2	1333.1663	129.4336	.2	1604.5999	142.0000
.3	1339.6458	129.7478	.3	1611.7077	142.3142
.4	1346.1410	130.0619	.4	1618.8313	142.6283
.5	1352.6520	130.3761	.5	1625.9705	142.9425
.6	1359.1786	130.6903	.6	1633.1255	143.2566
.7	1365.7210	131.0044	.7	1640.2962	143.5708
.8	1372.2791	131.3186	.8	1647.4826	143.8849
.9	1378.8529	131.6327	.9	1654.6847	144.1991
42.0	1385.4424	131.9469	46.0	1661.9025	144.5133
.1	1392.0476	132.2611	.1	1669.1360	144.8274
.2	1398.6685	132.5752	.2	1676.3853	145.1416
.3	1405.3051	132.8894	.3	1683.6502	145.4557
.4	1411.9574	133.2035	.4	1690.9308	145.7699
.5	1418.6254	133.5177	.5	1698.2272	146.0841
.6	1425.3092	133.8318	.6	1705.5392	146.3982
.7	1432.0086	134.1460	.7	1712.8670	146.7124
.8	1438.7238	134.4602	.8	1720.2105	147.0265
.9	1445.4546	134.7743	.9	1727.5697	147.3407
43.0	1452.2012	135.0885	47.0	1734.9445	147.6550
.1	1458.9635	135.4026	.1	1742.3351	147.9690
.2	1465.7415	135.7168	.2	1749.7414	148.2832
.3	1472.5352	136.0310	.3	1757.1635	148.5973
.4	1479.3446	136.3451	.4	1764.6012	148.9115
.5	1486.1697	136.6593	.5	1772.0546	149.2257
.6	1493.0105	136.9734	.6	1779.5237	149.5398
.7	1499.8670	137.2876	.7	1787.0086	149.8540
.8	1506.7393	137.6018	.8	1794.5091	150.1681
.9	1513.6272	137.9159	.9	1802.0254	150.4823

AREAS and CIRCUMFERENCES OF CIRCLES.
(CONTINUED.)

Diam.	Area.	Circum.	Diam.	Area.	Circum.
48.0	1809.5574	150.7964	52.0	2123.7166	163.3628
.1	1817.1050	151.1106	.1	2131.8926	163.6770
.2	1824.6684	151.4248	.2	2140.0843	163.9911
.3	1832.2475	151.7389	.3	2148.2917	164.3053
.4	1839.8423	152.0531	.4	2156.5149	164.6195
.5	1847.4528	152.3672	.5	2164.7537	164.9336
.6	1855.0790	152.6814	.6	2173.0082	165.2479
.7	1862.7210	152.9956	.7	2181.2785	165.5619
.8	1870.3786	153.3097	.8	2189.5644	165.8761
.9	1878.0519	153.6239	.9	2197.8661	166.1903
49.0	1885.7409	153.9380	53.0	2206.1834	166.5044
.1	1893.4457	154.2522	.1	2214.5165	166.8186
.2	1901.1662	154.5664	.2	2222.8653	167.1327
.3	1908.9024	154.8805	.3	2231.2298	167.4469
.4	1916.6543	155.1947	.4	2239.6100	167.7610
.5	1924.4218	155.5088	.5	2248.0059	168.0752
.6	1932.2051	155.8230	.6	2256.4175	168.3894
.7	1940.0042	156.1372	.7	2264.8448	168.7035
.8	1947.8189	156.4513	.8	2273.2879	169.0177
.9	1955.6493	156.7655	.9	2281.7466	169.3318
50.0	1963.4954	157.0796	54.0	2290.2210	169.6460
.1	1971.3572	157.3938	.1	2298.7112	169.9602
.2	1979.2348	157.7080	.2	2307.2171	170.2743
.3	1987.1280	158.0221	.3	2315.7386	170.5885
.4	1995.0370	158.3363	.4	2324.2759	170.9026
.5	2002.9617	158.6504	.5	2332.8289	171.2168
.6	2010.9020	158.9646	.6	2341.3976	171.5310
.7	2018.8581	159.2787	.7	2349.9820	171.8451
.8	2026.8299	159.5929	.8	2358.5821	172.1593
.9	2034.8174	159.9071	.9	2367.1979	172.4735
51.0	2042.8206	160.2212	55.0	2375.8294	172.7876
.1	2050.8395	160.5354	.1	2384.4767	173.1017
.2	2058.8742	160.8495	.2	2393.1396	173.4159
.3	2066.9245	161.1637	.3	2401.8183	173.7301
.4	2074.9905	161.4779	.4	2410.5126	174.0442
.5	2083.0723	161.7920	.5	2419.2227	174.3584
.6	2091.1697	162.1062	.6	2427.9485	174.6726
.7	2099.2829	162.4203	.7	2436.6899	174.9867
.8	2107.4118	162.7345	.8	2445.4471	175.3009
.9	2115.5563	163.0487	.9	2454.2200	175.6150

AREAS and CIRCUMFERENCES OF CIRCLES.
(CONTINUED.)

Diam.	Area.	Circum.	Diam.	Area.	Circum.
56.0	2463.0086	175.9292	60.0	2827.4334	188.4956
.1	2471.8130	176.2433	.1	2836.8660	188.8097
.2	2480.6330	176.5575	.2	2846.3144	189.1239
.3	2489.4687	176.8717	.3	2855.7784	189.4380
.4	2498.3201	177.1858	.4	2865.2582	189.7522
.5	2507.1873	177.5000	.5	2874.7536	190.0664
.6	2516.0701	177.8141	.6	2884.2648	190.3805
.7	2524.9687	178.1283	.7	2893.7917	190.6947
.8	2533.8830	178.4425	.8	2903.3343	191.0088
.9	2542.8129	178.7566	.9	2912.8926	191.3230
57.0	2551.7586	179.0708	61.0	2922.4666	191.6372
.1	2560.7200	179.3849	.1	2932.0563	191.9513
.2	2569.6971	179.6991	.2	2941.6617	192.2655
.3	2578.6899	180.0133	.3	2951.2828	192.5796
.4	2587.6985	180.3274	.4	2960.9197	192.8938
.5	2596.7227	180.6416	.5	2970.5722	193.2079
.6	2605.7626	180.9557	.6	2980.2405	193.5221
.7	2614.8183	181.2699	.7	2989.9244	193.8363
.8	2623.8896	181.5841	.8	2999.6241	194.1504
.9	2632.9767	181.8982	.9	3009.3395	194.4646
58.0	2642.0794	182.2124	62.0	3019.0705	194.7787
.1	2651.1979	182.5265	.1	3028.8173	195.0929
.2	2660.3321	182.8407	.2	3038.5798	195.4071
.3	2669.4820	183.1549	.3	3048.3580	195.7212
.4	2678.6476	183.4690	.4	3058.1520	196.0354
.5	2687.8289	183.7832	.5	3067.9616	196.3495
.6	2697.0259	184.0973	.6	3077.7869	196.6637
.7	2706.2386	184.4115	.7	3087.6279	196.9779
.8	2715.4670	184.7256	.8	3097.4847	197.2920
.9	2724.7112	185.0398	.9	3107.3571	197.6062
59.0	2733.9710	185.3540	63.0	3117.2453	197.9203
.1	2743.2466	185.6681	.1	3127.1492	198.2345
.2	2752.5378	185.9823	.2	3137.0688	198.5487
.3	2761.8448	186.2964	.3	3147.0040	198.8628
.4	2771.1675	186.6106	.4	3156.9550	199.1770
.5	2780.5058	186.9248	.5	3166.9217	199.4911
.6	2789.8599	187.2389	.6	3176.9043	199.8053
.7	2799.2297	187.5531	.7	3186.9023	200.1195
.8	2808.6152	187.8672	.8	3196.9161	200.4336
.9	2818.0165	188.1814	.9	3206.9456	200.7478

AREAS and CIRCUMFERENCES OF CIRCLES.
(CONTINUED.)

Diam.	Area.	Circum.	Diam.	Area.	Circum.
64.0	3216.9909	201.0620	68.0	3631.6811	213.6283
.1	3227.0518	201.3761	.1	3642.3704	213.9425
.2	3237.1285	201.6902	.2	3653.0754	214.2566
.3	3247.2222	202.0044	.3	3663.7960	214.5708
.4	3257.3289	202.3186	.4	3674.5324	214.8849
.5	3267.4527	202.6327	.5	3685.2845	215.1991
.6	3277.5922	202.9469	.6	3696.0523	215.5133
.7	3287.7474	203.2610	.7	3706.8359	215.8274
.8	3297.9183	203.5752	.8	3717.6351	216.1416
.9	3308.1049	203.8894	.9	3728.4500	216.4556
65.0	3318.3072	204.2035	69.0	3739.2807	216.7699
.1	3328.5253	204.5176	.1	3750.1270	217.0841
.2	3338.7590	204.8318	.2	3760.9891	217.3982
.3	3349.0085	205.1460	.3	3771.8668	217.7124
.4	3359.2736	205.4602	.4	3782.7603	218.0265
.5	3369.5545	205.7743	.5	3793.6695	218.3407
.6	3379.8510	206.0885	.6	3804.5944	218.6548
.7	3390.1633	206.4026	.7	3815.5350	218.9690
.8	3400.4913	206.7168	.8	3826.4913	219.2832
.9	3410.8350	207.0310	.9	3837.4633	219.5973
66.0	3421.1944	207.3451	70.0	3848.4510	219.9115
.1	3431.5695	207.6593	.1	3859.4544	220.2256
.2	3441.9603	207.9734	.2	3870.4736	220.5398
.3	3452.3669	208.2876	.3	3881.5084	220.8540
.4	3462.7891	208.6017	.4	3892.5590	221.1681
.5	3473.2270	208.9159	.5	3903.6252	221.4823
.6	3483.6807	209.2301	.6	3914.7072	221.7964
.7	3494.1500	209.5442	.7	3925.8049	222.1106
.8	3504.6351	209.8584	.8	3936.9182	222.4248
.9	3515.1359	210.1725	.9	3948.0473	222.7389
67.0	3525.6524	210.4867	71.0	3959.1921	223.0531
.1	3536.1845	210.8009	.1	3970.3526	223.3672
.2	3546.7324	211.1150	.2	3981.5289	223.6814
.3	3557.2960	211.4292	.3	3992.7208	223.9956
.4	3567.8754	211.7433	.4	4003.9284	224.3097
.5	3578.4704	212.0575	.5	4015.1518	224.6239
.6	3589.0811	212.3717	.6	4026.3908	224.9380
.7	3599.7075	212.6858	.7	4037.6456	225.2522
.8	3610.3497	213.0000	.8	4048.9160	225.5664
.9	3621.0075	213.3141	.9	4060.2022	225.8805

AREAS and CIRCUMFERENCES OF CIRCLES.
(CONTINUED.)

Diam.	Area.	Circum.	Diam.	Area.	Circum.
72.0	4071.5041	226.1947	76.0	4536.4598	238.7610
.1	4082.8217	226.5088	.1	4548.4057	239.0752
.2	4094.1550	226.8230	.2	4560.3673	239.3894
.3	4105.5040	227.1371	.3	4572.3446	239.7035
.4	4116.8687	227.4513	.4	4584.3377	240.0177
.5	4128.2491	227.7655	.5	4596.3464	240.3318
.6	4139.6452	228.0796	.6	4608.3708	240.6460
.7	4151.0571	228.3938	.7	4620.4110	240.9602
.8	4162.4846	228.7079	.8	4632.4669	241.2743
.9	4173.9279	229.0221	.9	4644.5384	241.5885
73.0	4185.3868	229.3363	77.0	4656.6257	241.9026
.1	4196.8615	229.6504	.1	4668.7287	242.2168
.2	4208.3519	229.9646	.2	4680.8474	242.5310
.3	4219.8579	230.2787	.3	4692.9818	242.8451
.4	4231.3797	230.5929	.4	4705.1319	243.1592
.5	4242.9172	230.9071	.5	4717.2977	243.4734
.6	4254.4704	231.2212	.6	4729.4792	243.7876
.7	4266.0394	231.5354	.7	4741.6765	244.1017
.8	4277.6240	231.8495	.8	4753.8894	244.4159
.9	4289.2243	232.1637	.9	4766.1181	244.7301
74.0	4300.8403	232.4779	78.0	4778.3624	245.0442
.1	4312.4721	232.7920	.1	4790.6225	245.3584
.2	4324.1195	233.1062	.2	4802.8983	245.6725
.3	4335.7827	233.4203	.3	4815.1897	245.9867
.4	4347.4616	233.7345	.4	4827.4969	246.3009
.5	4359.1562	234.0487	.5	4839.8198	246.6150
.6	4370.8664	234.3628	.6	4852.1584	246.9292
.7	4382.5924	234.6770	.7	4864.5128	247.2433
.8	4394.3341	234.9911	.8	4876.8828	247.5575
.9	4406.0916	235.3053	.9	4889.2685	247.8717
75.0	4417.8647	235.6194	79.0	4901.6699	248.1858
.1	4429.6535	235.9336	.1	4914.0871	248.5000
.2	4441.4580	236.2478	.2	4926.5199	248.8141
.3	4453.2783	236.5619	.3	4938.9685	249.1283
.4	4465.1142	236.8761	.4	4951.4328	249.4425
.5	4476.9659	237.1902	.5	4963.9127	249.7566
.6	4488.8332	237.5044	.6	4976.4084	250.0708
.7	4500.7163	237.8186	.7	4988.9198	250.3850
.8	4512.6151	238.1327	.8	5001.4469	250.6991
.9	4524.5296	238.4469	.9	5013.9897	251.0133

AREAS and CIRCUMFERENCES OF CIRCLES.
(CONTINUED.)

Diam.	Area.	Circum.	Diam.	Area.	Circum.
80.0	5026.5482	251.3274	84.0	5541.7694	263.8938
.1	5039.1225	251.6416	.1	5554.9720	264.2079
.2	5051.7124	251.9557	.2	5568.1902	264.5221
.3	5064.3180	252.2699	.3	5581.4242	264.8363
.4	5076.9394	252.5840	.4	5594.6739	265.1514
.5	5089.5764	252.8982	.5	5607.9392	265.4646
.6	5102.2292	253.2124	.6	5621.2203	265.7787
.7	5114.8977	253.5265	.7	5634.5171	266.0929
.8	5127.5819	253.8407	.8	5647.8296	266.4071
.9	5140.2818	254.1548	.9	5661.1578	266.7212
81.0	5152.9973	254.4690	85.0	5674.5017	267.0354
.1	5165.7287	254.7832	.1	5687.8614	267.3495
.2	5178.4757	255.0973	.2	5701.2367	267.6637
.3	5191.2384	255.4115	.3	5714.6277	267.9779
.4	5204.0168	255.7256	.4	5728.0345	268.2920
.5	5216.8110	256.0398	.5	5741.4569	268.6062
.6	5229.6208	256.3540	.6	5754.8951	268.9203
.7	5242.4463	256.6681	.7	5768.3490	269.2345
.8	5255.2876	256.9823	.8	5781.8185	269.5486
.9	5268.1446	257.2966	.9	5795.3038	269.8628
82.0	5281.0173	257.6106	86.0	5808.8048	270.1770
.1	5293.9056	257.9247	.1	5822.3215	270.4911
.2	5306.8097	258.2389	.2	5835.8539	270.8053
.3	5319.7295	258.5531	.3	5849.4020	271.1194
.4	5332.6650	258.8672	.4	5862.9659	271.4336
.5	5345.6162	259.1814	.5	5876.5454	271.7478
.6	5358.5832	259.4956	.6	5890.1407	272.0619
.7	5371.5658	259.8097	.7	5903.7516	272.3761
.8	5384.5641	260.1239	.8	5917.3783	272.6902
.9	5397.5782	260.4380	.9	5931.0206	273.0044
83.0	5410.6079	260.7522	87.0	5944.6787	273.3186
.1	5423.6534	261.0663	.1	5958.3525	273.6327
.2	5436.7146	261.3805	.2	5972.0420	273.9469
.3	5449.7915	261.6947	.3	5985.7472	274.2610
.4	5462.8840	262.0088	.4	5999.4681	274.5752
.5	5475.9923	262.3230	.5	6013.2047	274.8894
.6	5489.1163	262.6371	.6	6026.9570	275.2035
.7	5502.2561	262.9513	.7	6040.7250	275.5177
.8	5515.4115	263.2655	.8	6054.5088	275.8318
.9	5528.5826	263.5796	.9	6068.3082	276.1460

AREAS and CIRCUMFERENCES OF CIRCLES.
(CONTINUED.)

Diam.	Area.	Circum.	Diam.	Area.	Circum.
88.0	6082.1234	276.4602	92.0	6647.6101	289.0265
.1	6095.9542	276.7743	.1	6662.0692	289.3407
.2	6109.8008	277.0885	.2	6676.5441	289.6548
.3	6123.6631	277.4026	.3	6691.0347	289.9690
.4	6137.5411	277.7168	.4	6705.5410	290.2832
.5	6151.4348	278.0309	.5	6720.0630	290.5973
.6	6165.3442	278.3451	.6	6734.6008	290.9115
.7	6179.2693	278.6593	.7	6749.1542	291.2256
.8	6193.2101	278.9740	.8	6763.7233	291.5398
.9	6207.1666	279.2876	.9	6778.3082	291.8540
89.0	6221.1389	279.6017	93.0	6792.9087	292.1681
.1	6235.1268	279.9159	.1	6807.5250	292.4823
.2	6249.1304	280.2301	.2	6822.1569	292.7964
.3	6263.1498	280.5442	.3	6836.8046	293.1106
.4	6277.1849	280.8584	.4	6851.4680	293.4248
.5	6291.2356	281.1725	.5	6866.1471	293.7389
.6	6305.3021	281.4867	.6	6880.8419	294.0531
.7	6319.3843	281.8009	.7	6895.5524	294.3672
.8	6333.4822	282.1150	.8	6910.2786	294.6814
.9	6347.5958	282.4292	.9	6925.0205	294.9956
90.0	6361.7251	282.7433	94.0	6939.7782	295.3097
.1	6375.8701	283.0575	.1	6954.5515	295.6239
.2	6390.0309	283.3717	.2	6969.3106	295.9380
.3	6404.2073	283.6858	.3	6984.1453	296.2522
.4	6418.3995	284.0000	.4	6998.9658	296.5663
.5	6432.6073	284.3141	.5	7013.8019	296.8805
.6	6446.8309	284.6283	.6	7028.6538	297.1947
.7	6461.0701	284.9425	.7	7043.5214	297.5088
.8	6475.3251	285.2566	.8	7058.4047	297.8230
.9	6489.5958	285.5708	.9	7073.3033	298.1371
91.0	6503.8822	285.8849	95.0	7088.2184	298.4513
.1	6518.1843	286.1991	.1	7103.1488	298.7655
.2	6532.5021	286.5133	.2	7118.1950	299.0796
.3	6546.8356	286.8274	.3	7133.0568	299.3938
.4	6561.1848	287.1416	.4	7148.0343	299.7079
.5	6575.5498	287.4557	.5	7163.0276	300.0221
.6	6589.9304	287.7699	.6	7178.0366	300.3363
.7	6604.3268	288.0840	.7	7193.0612	300.6504
.8	6618.7388	288.3982	.8	7208.1016	300.9646
.9	6633.1666	288.7124	.9	7223.1577	301.2787

AREAS and CIRCUMFERENCES OF CIRCLES.
(CONTINUED.)

Diam.	Area.	Circum.	Diam.	Area.	Circum.
96.0	7238.2295	301.5929	98.0	7542.9640	307.8761
.1	7253.3170	301.9071	.1	7558.3656	308.1902
.2	7268.4202	302.2212	.2	7573.7830	308.5044
.3	7283.5391	302.5354	.3	7589.2161	308.8186
.4	7298.6737	302.8405	.4	7604.6648	309.1327
.5	7313.8240	303.1637	.5	7620.1293	309.4469
.6	7328.9901	303.4779	.6	7635.6095	309.7610
.7	7344.1718	303.7920	.7	7651.1054	310.0752
.8	7359.3693	304.1062	.8	7666.6170	310.3894
.9	7374.5824	304.4203	.9	7682.1444	310.7035
97.0	7389.8113	304.7345	99.0	7697.6893	311.0177
.1	7405.0559	305.0486	.1	7713.2461	311.3318
.2	7420.3162	305.3628	.2	7728.8206	311.6460
.3	7435.5922	305.6770	.3	7744.4107	311.9602
.4	7450.8839	305.9911	.4	7760.0166	312.2743
.5	7466.1913	306.3053	.5	7775.6382	312.5885
.6	7481.5144	306.6194	.6	7791.2754	312.9026
.7	7496.8532	306.9336	.7	7806.9284	313.2168
.8	7512.2078	307.2478	.8	7822.5971	313.5309
.9	7527.5780	307.5619	.9	7838.2815	313.8451
			100.0	7853.9816	314.1593

To compute the area or circumference of a diameter greater than 100 and less than 1001:

Take out the area or circumference from table as though the number had one decimal, and move the decimal point two places to the right for the area, and one place for the circumference.

EXAMPLE—Wanted the area and circumference of 567. The tabular area for 56.7 is 2524.9687; and circumference 178.1283. Therefore area of 567 = 252496.87 and circumference = 1781.283.

To compute the area or circumference of a diameter greater than 1000:

Divide by a factor, as 2, 3, 4, 5, etc., if practicable, that will leave a quotient to be found in table, then multiply the tabular area of the quotient by the *square* of the factor, or the tabular circumference by the factor.

EXAMPLE—Wanted the area and circumference of 2109. Dividing by 3, the quotient is 703, for which the area is 388150.84 and the circumference 2208.54. Therefore area of 2109 = 388150.84 × 9 = 3493357.56 and circumference = 2208.54 × 3 = 6625.62.

WEIGHT OF RIVETS, and ROUND HEADED BOLTS WITHOUT NUTS, PER 100.

Length from under head.   One cubic foot weighing 480 lbs.

Length. Inches.	3/8" Dia.	1/2" Dia.	5/8" Dia.	3/4" Dia.	7/8" Dia.	1" Dia.	1 1/8" Dia.	1 1/4" Dia.
1 1/4	5.4	12.6	21.5	28.7	43.1	65.3	91.5	123.
1 1/2	6.2	13.9	23.7	31.8	47.3	70.7	98.4	133.
1 3/4	6.9	15.3	25.8	34.9	51.4	76.2	105.	142.
2	7.7	16.6	27.9	37.9	55.6	81.6	112.	150.
2 1/4	8.5	18.0	30.0	41.0	59.8	87.1	119.	159.
2 1/2	9.2	19.4	32.2	44.1	63.0	92.5	126.	167.
2 3/4	10.0	20.7	34.3	47.1	68.1	98.0	133.	176.
3	10.8	22.1	36.4	50.2	72.3	103.	140.	184.
3 1/4	11.5	23.5	38.6	53.3	76.5	109.	147.	193.
3 1/2	12.3	24.8	40.7	56.4	80.7	114.	154.	201.
3 3/4	13.1	26.2	42.8	59.4	84.8	120.	161.	210.
4	13.8	27.5	45.0	62.5	89.0	125.	167.	218.
4 1/4	14.6	28.9	47.1	65.6	93.2	131.	174.	227.
4 1/2	15.4	30.3	49.2	68.6	97.4	136.	181.	236.
4 3/4	16.2	31.6	51.4	71.7	102.	142.	188.	244.
5	16.9	33.0	53.5	74.8	106.	147.	195.	253.
5 1/4	17.7	34.4	55.6	77.8	110.	153.	202.	261.
5 1/2	18.4	35.7	57.7	80.9	114.	158.	209.	270.
5 3/4	19.2	37.1	59.9	84.0	118.	163.	216.	278.
6	20.0	38.5	62.0	87.0	122.	169.	223.	287.
6 1/2	21.5	41.2	66.3	93.2	131.	180.	236.	304.
7	23.0	43.9	70.5	99.3	139.	191.	250.	321.
7 1/2	24.6	46.6	74.8	106.	147.	202.	264.	338.
8	26.1	49.4	79.0	112.	156.	213.	278.	355.
8 1/2	27.6	52.1	83.3	118.	164.	223.	292.	372.
9	29.2	54.8	87.6	124.	173.	234.	306.	389.
9 1/2	30.7	57.6	91.8	130.	181.	245.	319.	406.
10	32.2	60.3	96.1	136.	189.	256.	333.	423.
10 1/2	33.8	63.0	101.	142.	198.	267.	347.	440.
11	35.3	65.7	105.	148.	206.	278.	361.	457.
11 1/2	36.8	68.5	109.	155.	214.	289.	375.	474.
12	38.4	71.2	113.	161.	223.	300.	388.	491.
Heads.	1.8	5.7	10.9	13.4	22.2	38.0	57.0	82.0

UPSET SCREW ENDS FOR ROUND AND SQUARE BARS.

Standard Proportions of the Keystone Bridge Company.

Dia. of Round or Side of Square Bar. Inches.	ROUND BARS.				SQUARE BARS.			
	Dia. of Upset Screw End. Inches.	Dia. of Screw at Root of Thread. Inches.	Threads per Inch. No.	Excess of Effective Area of Screw End over Bar. Per Cent.	Dia. of Upset Screw End. Inches.	Dia. of Screw at Root of Thread. Inches.	Threads per Inch. No.	Excess of Effective Area of Screw End over Bar. Per Cent.
½	¾	.620	10	54	¾	.620	10	21
9/16	¾	.620	10	21	⅞	.731	9	33
⅝	⅞	.731	9	37	1	.837	8	41
11/16	1	.837	8	48	1	.837	8	17
¾	1	.837	8	25	1⅛	.940	7	23
13/16	1⅛	.940	7	34	1¼	1.065	7	35
⅞	1¼	1.065	7	48	1⅜	1.160	6	38
15/16	1¼	1.065	7	29	1⅜	1.160	6	20
1	1⅜	1.160	6	35	1½	1.284	6	29
1 1/16	1⅜	1.160	6	19	1⅝	1.389	5½	34
1⅛	1½	1.284	6	30	1⅝	1.389	5½	20
1 3/16	1½	1.284	6	17	1¾	1.490	5	24
1¼	1⅝	1.389	5½	23	1⅞	1.615	5	31
1 5/16	1¾	1.490	5	29	1⅞	1.615	5	19
1⅜	1¾	1.490	5	18	2	1.712	4½	22
1 7/16	1⅞	1.615	5	26	2⅛	1.837	4½	28
1½	2	1.712	4½	30	2⅛	1.837	4½	18
1 9/16	2	1.712	4½	20	2¼	1.962	4½	24
1⅝	2⅛	1.837	4½	28	2⅜	2.087	4½	30
1 11/16	2⅛	1.837	4½	18	2⅜	2.087	4½	20
1¾	2¼	1.962	4½	26	2½	2.175	4	21
1 13/16	2¼	1.962	4½	17	2⅝	2.300	4	26
1⅞	2⅜	2.087	4½	24	2⅝	2.300	4	18
1 15/16	2½	2.175	4	26	2¾	2.425	4	23
2	2½	2.175	4	18	2⅞	2.550	4	28
2 1/16	2⅝	2.300	4	24	2⅞	2.550	4	20
2⅛	2⅝	2.300	4	17	3	2.629	3½	20
2 3/16	2¾	2.425	4	23	3⅛	2.754	3½	24

UPSET SCREW ENDS.

(CONTINUED.)

Dia. of Round or Side of Square Bar. Inches.	ROUND BARS.				SQUARE BARS.			
	Dia. of Upset Screw End. Inches.	Dia. of Screw at Root of Thread. Inches.	Threads per Inch. No.	Excess of Effective Area of Screw End over Bar. Per Cent.	Dia. of Upset Screw End. Inches.	Dia. of Screw at Root of Thread. Inches.	Threads per Inch. No.	Excess of Effective Area of Screw End over Bar. Per Cent.
2¼	2⅞	2.550	4	28	3⅛	2.754	3½	18
2 5/16	2⅞	2.550	4	22	3¼	2.879	3½	22
2⅜	3	2.629	3½	23	3⅜	3.004	3½	26
2 7/16	3⅛	2.754	3½	28	3⅜	3.004	3½	19
2½	3⅛	2.754	3½	21	3½	3.100	3¼	21
2 9/16	3¼	2.879	3½	26	3⅝	3.225	3¼	24
2⅝	3¼	2.879	3½	20	3⅝	3.225	3¼	19
2 11/16	3⅜	3.004	3½	25	3¾	3.317	3	20
2¾	3⅜	3.004	3½	19	3⅞	3.442	3	23
2 13/16	3½	3.100	3¼	22	3⅞	3.442	3	18
2⅞	3⅝	3.225	3¼	26	4	3.567	3	21
2 15/16	3⅝	3.225	3¼	21	4⅛	3.692	3	24
3	3¾	3.317	3	22	4⅛	3.692	3	19
3⅛	3⅞	3.442	3	21	4⅜	3.923	2⅞	24
3¼	4	3.567	3	20	4½	4.028	2¾	21
3⅜	4⅛	3.692	3	20	4⅝	4.153	2¾	19
3½	4¼	3.798	2⅞	18				
3⅝	4½	4.028	2¾	23				
3¾	4⅝	4.153	2¾	23				
3⅞	4¾	4.255	2⅝	21				

REMARKS.—As upsetting reduces the strength of iron, bars having the same diameter at root of thread as that of the bar, invariably break in the screw end, when tested to destruction, without developing the full strength of the bar. It is therefore necessary to make up for this loss in strength by an excess of metal in the upset screw ends over that in the bar.

The above table is the result of numerous tests on finished bars made at the Keystone Bridge Company's Works in Pittsburgh, and gives proportions that will cause the bar to break in the body in preference to the upset end.

The screw threads in above table are the Franklin Institute standard.

To make one upset end for 5″ length of thread allow 6″ length of rod additional.

STANDARD SCREW THREADS, NUTS AND BOLT HEADS.—Recommended by the Franklin Institute.

SCREW THREADS.

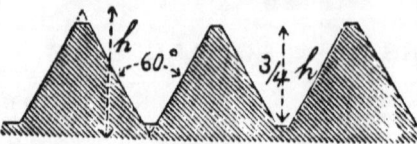

Angle of Thread 60°. Flat at Top and Bottom = 1/8 of pitch.

Dia. of Screw. Inches.	Dia. at Root of Thread. Inches.	Threads per Inch. No.
¼	.185	20
5/16	.240	18
⅜	.294	16
7/16	.344	14
½	.400	13
9/16	.454	12
⅝	.507	11
¾	.620	10
⅞	.731	9
1	.837	8
1⅛	.940	7
1¼	1.065	7
1⅜	1.160	6
1½	1.284	6
1⅝	1.389	5½
1¾	1.490	5
1⅞	1.615	5
2	1.712	4½
2¼	1.962	4½
2½	2.175	4
2¾	2.425	4
3	2.629	3½
3¼	2.879	3½
3½	3.100	3¼
3¾	3.317	3
4	3.567	3
4¼	3.798	2⅞
4½	4.028	2¾
4¾	4.255	2⅝
5	4.480	2½
5¼	4.730	2½
5½	5.053	2⅜
5¾	5.203	2⅜
6	5.423	2¼

Nuts and Bolt Heads are determined by the following rules, which apply to Square and Hexagon Nuts both:

Short diameter of rough nut = 1½ × dia. of bolt + ⅛ in.

Short diameter of finished nut = 1½ × dia. of bolt + 1-16 in.

Thickness of rough nut = diameter of bolt.

Thickness of finished nut = diameter of bolt − 1-16 in.

Short diameter of rough head = 1½ × dia. of bolt + ⅛ in.

Short dia. of finished head = 1½ × dia. of bolt + 1-16 in.

Thickness of rough head = ½ short dia. of head.

Thickness of finished head = dia. of bolt − 1-16 in.

The long diameter of a hexagon nut may be obtained by multiplying the short diameter by 1.155, and the long diameter of a square nut by multiplying the short diameter by 1.414.

The above standards for screw threads, nuts and bolt heads, were recommended by the Franklin Institute in Dec. 1864. The standard for screw threads has been very generally adopted in the United States, but the proportions recommended for nuts and bolt heads have not found general acceptance because of the odd sizes of bar —not usually rolled by the mills—which they would require from which to make the nut.

WHITWORTH'S STANDARD ANGULAR SCREW THREADS.

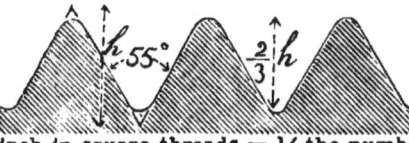

Angle of Thread 55°.
Depth of Thread = pitch of screw.
⅙ of depth is rounded off at top and bottom.
Number of threads to the inch in square threads = ½ the number in angular threads.

Dia. of Screw. In.	Threads to the Inch. No.	Dia. of Screw. In.	Threads to the Inch. No.	Dia. of Screw. In.	Threads to the Inch. No.	Dia. of Screw. In.	Threads to the Inch. No.
1-4	20	1	8	2	4 1-2	4	3
5-16	18	1 1-8	7	2 1-4	4	4 1-4	2 7-8
3-8	16	1 1-4	7	2 1-2	4	4 1-2	2 7-8
7-16	14	1 3-8	6	2 3-4	3 1-2	4 3-4	2 3-4
1-2	12	1 1-2	6	3	3 1-2	5	2 3-4
5-8	11	1 5-8	5	3 1-4	3 1-4	5 1-4	2 5-8
3-4	10	1 3-4	5	3 1-2	3 1-4	5 1-2	2 5-8
7-8	9	1 7-8	4 1-2	3 3-4	3	5 3-4	2 1-2
						6	2 1-2

WOOD SCREWS.

Diameter = number × 0.01325 + 0.056.

No.	Dia.	No.	Dia.	No.	Dia.	No.	Dia.
0	.056	6	.135	12	.215	18	.293
1	.069	7	.149	13	.228	19	.308
2	.082	8	.162	14	.241	20	.321
3	.096	9	.175	15	.255	21	.334
4	.109	10	.188	16	.268	22	.347
5	.122	11	.201	17	.281	23	.361

No.	Dia.
24	.374
25	.387
26	.401
27	.414
28	.427
29	.440
30	.453

TACKS.

Title. Oz.	Length. In.	No. per lb.	Title. Oz.	Length. In.	No. per lb.	Title. Oz.	Length. In.	No. per lb.	Title. Oz.	Length. In.	No. per lb.
1	1-8	16000	3	3-8	5333	10	11-16	1600	18	15-16	888
1 1-2	3-16	10666	4	7-16	4000	12	3-4	1333	20	1	800
2	1-4	8000	6	9-16	2666	14	13-16	1143	22	1 1-16	727
2 1-2	5-16	6400	8	5-8	2000	16	7-8	1000	24	1 1-8	666

WROUGHT SPIKES.

Number to a keg of 150 lbs.

Length. In.	¼ in. No.	5/16 in. No.	3/8 in. No.	Length. In.	¼ in. No.	5/16 in. No.	3/8 in. No.	7/16 in. No.	½ in. No.
3	2250	......	......	7	1161	662	482	445	306
3 1-2	1890	1208	......	8	......	635	455	384	256
4	1650	1135	......	9	......	573	424	300	240
4 1-2	1464	1064	......	10	......	......	391	270	222
5	1380	930	742	11	......	......	......	249	203
6	1292	868	570	12	......	......	......	236	180

SIZES AND WEIGHTS OF HOT PRESSED SQUARE NUTS.

As manufactured by Charles & McMurtry, Pittsburgh, Pa. The sizes are the usual manufacturers', not the Franklin Institute Standard. Both weights and sizes are for the unfinished Nut.

Size of Bolt.	Weight of One Nut.	Rough Hole.	Thickness of Nut.	Side of Square.	Diagonal.	No. of Nuts in 100 lbs.
1/4	.014	7/32	1/4	1/2	.71	6900
5/16	.029	9/32	5/16	5/8	.88	3450
3/8	.048	11/32	3/8	3/4	1.06	2080
7/16	.078	13/32	7/16	7/8	1.24	1280
1/2	.088	7/16	1/2	7/8	1.24	1140
1/2	.116	7/16	1/2	1	1.41	860
9/16	.161	1/2	9/16	1 1/8	1.59	620
5/8	.172	9/16	5/8	1 1/8	1.59	580
5/8	.22	9/16	5/8	1 1/4	1.77	460
3/4	.31	21/32	3/4	1 3/8	1.94	320
3/4	.38	21/32	3/4	1 1/2	2.12	260
7/8	.56	25/32	7/8	1 5/8	2.30	180
7/8	.63	25/32	7/8	1 3/4	2.47	160
1	.69	7/8	1	1 3/4	2.47	144
1	.91	7/8	1	2	2.83	110
1 1/8	1.00	15/16	1 1/8	2	2.83	100
1 1/8	1.43	15/16	1 1/8	2 1/4	3.18	70
1 1/4	1.54	1 1/16	1 1/4	2 1/4	3.18	65
1 1/4	1.79	1 1/16	1 1/4	2 1/2	3.54	56
1 3/8	2.4	1 3/16	1 3/8	2 3/4	3.89	42
1 1/2	3.1	1 5/16	1 1/2	3	4.24	32
1 5/8	4.0	1 7/16	1 5/8	3 1/4	4.60	25
1 3/4	5.0	1 9/16	1 3/4	3 1/2	4.95	20
1 7/8	5.9	1 11/16	1 7/8	3 3/4	5.30	17
2	7.1	1 13/16	2	4	5.66	14.
2 1/8	7.4	1 7/8	2 1/8	4	5.66	13.5
2 1/4	8.1	2	2 1/4	4 1/4	6.01	12.3
2 3/8	8.3	2 1/8	2 3/8	4 1/4	6.01	12.0
2 1/2	10.9	2 1/4	2 1/2	4 1/2	6.36	9.14
2 3/4	13.2	2 7/16	2 3/4	4 3/4	6.72	7.55
3	14.9	2 11/16	3	5	7.07	6.72
3 1/4	17.5	2 15/16	3 1/4	5 1/2	7.78	5.70
3 1/2	21.1	3 1/2	3 1/2	6	8.49	4.75

SIZES AND WEIGHTS OF HOT PRESSED HEXAGON NUTS.

As manufactured by Charles & McMurtry, Pittsburgh, Pa. The sizes are the usual manufacturers', not the Franklin Institute Standard. Both weights and sizes are for the unfinished Nut.

Size of Bolt.	Weight of One Nut.	Rough Hole.	Thickness of Nut.	Short Diameter.	Long Diameter.	No. of Nuts in 100 lbs.
$\frac{1}{4}$	.013	$\frac{7}{32}$	$\frac{1}{4}$	$\frac{1}{2}$	.58	8000
$\frac{5}{16}$	.026	$\frac{9}{32}$	$\frac{5}{16}$	$\frac{5}{8}$	.72	3840
$\frac{3}{8}$	.042	$\frac{11}{32}$	$\frac{3}{8}$	$\frac{3}{4}$	.87	2400
$\frac{7}{16}$	.071	$\frac{13}{32}$	$\frac{7}{16}$	$\frac{7}{8}$	1.01	1400
$\frac{1}{2}$	.069	$\frac{7}{16}$	$\frac{1}{2}$	$\frac{7}{8}$	1.01	1440
$\frac{1}{2}$	.100	$\frac{7}{16}$	$\frac{1}{2}$	1	1.15	1000
$\frac{9}{16}$	.161	$\frac{1}{2}$	$\frac{9}{16}$	$1\frac{1}{8}$	1.30	620
$\frac{5}{8}$	.147	$\frac{9}{16}$	$\frac{5}{8}$	$1\frac{1}{8}$	1.30	680
$\frac{5}{8}$	.200	$\frac{9}{16}$	$\frac{5}{8}$	$1\frac{1}{4}$	1.44	500
$\frac{5}{8}$	.23	$\frac{9}{16}$	$\frac{3}{4}$	$1\frac{1}{4}$	1.44	400
$\frac{3}{4}$	.26	$\frac{21}{32}$	$\frac{3}{4}$	$1\frac{3}{8}$	1.59	380
$\frac{3}{4}$	.33	$\frac{21}{32}$	$\frac{7}{8}$	$1\frac{1}{2}$	1.73	300
$\frac{7}{8}$	.45	$\frac{25}{32}$	$\frac{7}{8}$	$1\frac{5}{8}$	1.88	220
$\frac{7}{8}$	.53	$\frac{25}{32}$	1	$1\frac{5}{8}$	1.88	190
1	.59	$\frac{7}{8}$	1	$1\frac{3}{4}$	2.02	170
1	.63	$\frac{7}{8}$	$1\frac{1}{8}$	$1\frac{3}{4}$	2.02	160
$1\frac{1}{8}$	.95	$1\frac{5}{16}$	$1\frac{1}{4}$	2	2.31	105
$1\frac{1}{4}$	1.43	$1\frac{1}{16}$	$1\frac{3}{8}$	$2\frac{1}{4}$	2.60	70
$1\frac{3}{8}$	1.64	$1\frac{3}{16}$	$1\frac{1}{2}$	$2\frac{1}{2}$	2.89	61
$1\frac{1}{2}$	2.4	$1\frac{5}{16}$	$1\frac{5}{8}$	$2\frac{3}{4}$	3.18	42
$1\frac{5}{8}$	3.0	$1\frac{7}{16}$	$1\frac{3}{4}$	3	3.46	33
$1\frac{3}{4}$	3.7	$1\frac{9}{16}$	$1\frac{7}{8}$	$3\frac{1}{4}$	3.75	27
$1\frac{7}{8}$	4.8	$1\frac{11}{16}$	2	$3\frac{1}{2}$	4.04	21
2	4.5	$1\frac{13}{16}$	2	$3\frac{1}{2}$	4.04	22
$2\frac{1}{8}$	5.1	$1\frac{7}{8}$	$2\frac{1}{8}$	$3\frac{3}{4}$	4.33	19.5
$2\frac{1}{4}$	5.4	2	$2\frac{1}{4}$	$3\frac{3}{4}$	4.33	18.4
$2\frac{3}{8}$	6.3	$2\frac{1}{8}$	$2\frac{3}{8}$	4	4.62	15.84
$2\frac{1}{2}$	7.6	$2\frac{1}{4}$	$2\frac{1}{2}$	$4\frac{1}{4}$	4.91	13.11
$2\frac{3}{4}$	9.3	$2\frac{7}{16}$	$2\frac{3}{4}$	$4\frac{1}{2}$	5.20	10.80
3	11.8	$2\frac{11}{16}$	3	$4\frac{3}{4}$	5.48	8.46
$3\frac{1}{4}$	15.9	$2\frac{15}{16}$	$3\frac{1}{4}$	5	5.77	6.30
$3\frac{1}{2}$	23.8	$3\frac{1}{8}$	$3\frac{1}{2}$	$5\frac{1}{4}$	6.06	4.20

WROUGHT IRON WELDED TUBES, FOR GAS, STEAM, OR WATER.

1¼ inch and below, Butt Welded; 1½ inch and above, Lap Welded; Proved to 300 lbs. per square inch by Hydraulic pressure.

TABLE OF STANDARD DIMENSIONS, AS MANUFACTURED BY MORRIS, TASKER & CO., LIMITED.

Inside Diameter.	Actual Outside Diameter.	Thickness.	Actual Inside Diameter.	Internal Circumference.	External Circumference.	Lgth of Pipe per sq. foot of Inside Surface.	Lgth of Pipe per sq. foot of Outside Surface.	Internal Area.	External Area.	Length of Pipe containing One cubic foot.	Weight per foot of Length.	No. of Threads per inch of Screw.	Taper of Threads per inch of Screw.
Inch.	Inches.	Inches.	Inches.	Inches.	Inches.	Feet.	Feet.	Inches.	Inches.	Feet.	Lbs.		Inch.
⅛	0.405	0.068	0.270	0.848	1.272	14.15	9.44	0.0572	0.129	2500.	0.243	27	1/32
¼	0.54	0.088	0.364	1.144	1.696	10.50	7.075	0.1041	0.229	1385.	0.422	18	1/32
⅜	0.675	0.091	0.494	1.552	2.121	7.67	5.657	0.1916	0.358	751.5	0.561	18	1/32
½	0.84	0.109	0.623	1.957	2.652	6.13	4.502	0.3048	0.554	472.4	0.845	14	1/32
¾	1.05	0.113	0.824	2.589	3.299	4.635	3.637	0.5333	0.866	270.	1.126	14	1/32
1	1.315	0.134	1.048	3.292	4.134	3.679	2.903	0.8627	1.357	166.9	1.670	11½	1/32
1¼	1.66	0.140	1.380	4.335	5.215	2.768	2.301	1.496	2.164	96.25	2.258	11½	1/32
1½	1.9	0.145	1.611	5.061	5.969	2.371	2.01	2.038	2.835	70.65	2.694	11½	1/32
2	2.375	0.154	2.067	6.494	7.461	1.848	1.611	3.355	4.430	42.36	3.667	11½	1/32
2½	2.875	0.204	2.468	7.754	9.032	1.547	1.328	4.783	6.491	30.11	5.773	8	1/32
3	3.5	0.217	3.067	9.636	10.996	1.245	1.091	7.388	9.621	19.49	7.547	8	1/32
3½	4.0	0.226	3.548	11.146	12.566	1.077	0.955	9.887	12.566	14.56	9.055	8	1/32
4	4.5	0.237	4.026	12.648	14.137	0.949	0.849	12.730	15.904	11.31	10.728	8	1/32
4½	5.	0.247	4.508	14.153	15.708	0.848	0.765	15.939	19.635	9.03	12.492	8	1/32
5	5.563	0.259	5.045	15.849	17.475	0.757	0.629	19.990	24.299	7.20	14.564	8	1/32
6	6.625	0.280	6.065	19.054	20.813	0.63	0.577	28.889	34.471	4.98	18.767	8	1/32
7	7.625	0.301	7.023	22.063	23.954	0.544	0.505	38.737	45.663	3.72	23.410	8	1/32
8	8.625	0.322	7.982	25.076	27.096	0.478	0.444	50.039	58.426	2.88	28.348	8	1/32
9	9.688	0.344	9.001	28.277	30.433	0.425	0.394	63.633	73.715	2.26	34.077	8	1/64
10	10.75	0.366	10.019	31.475	33.772	0.381	0.355	78.838	90.762	1.80	40.641	8	1/64

EXPLANATION OF TABLES ON RIVETS AND PINS.

Pages 135 to 137, inclusive.

In transmitting stress by means of rivets, it is customary to disregard the friction between the parts joined, as too uncertain an element to be relied upon to any extent. The rivets must then be proportioned for the entire stress which is to be transmitted from one plate, or group of plates, to the other, and they must be of sufficient size and number, to present ample resistance to shearing and afford sufficient bearing area, so as not to cause a crushing of the metal at the rivet holes. This latter condition, while generally observed for pins, is very often entirely overlooked in riveted work. Its observance, in most cases of riveted girders with single webs, determines the size and number of rivets to be used, and frequently makes it necessary to adopt a greater thickness of web than would otherwise be required. Thus, if the web is $\frac{5}{16}''$ thick, the rivets connecting the same with the flange angles have a bearing value of only 3520 lbs. for a $\frac{3}{4}''$ rivet, while their shearing value is $= 2 \times 3310 = 6620$ lbs. per rivet, the rivets being in double shear. Consequently, while the usual thickness of web of floorbeams for railway bridges is $\frac{3}{8}''$, it sometimes becomes necessary, for shallow floorbeams, to increase this thickness to $\frac{1}{2}''$ and even $\frac{5}{8}''$, in order that the pressure of the rivets upon the semi-intrados of the rivet holes be not excessive, between the points of support of floorbeam and of application of the load, (in which space the transmission of stress from web to flanges takes place.)

The pressure usually allowed upon rivet-bearing is 15000 lbs. per square inch, as assumed in table, the bearing area being the diameter of hole multiplied by the thickness of metal. This pressure is somewhat greater than is generally allowed for pins, in consideration of the neglect of the friction between plates in riveted work.

Pins must be calculated for shearing, bending and bearing stresses, but one of the latter two only, in almost every case, determines the size to be used. The stress allowed upon pin-bearing in bridges proportioned to a factor of safety of five, is usually 12500 lbs., and the maximum fiber strain by bending, 15000 lbs. per square inch. Where groups of bars are connected to the same pin, as in the lower chords of truss bridges, the size of bars must be so chosen and the bars so placed that at no point on the pin will there be an excessive bending strain, on the presumption that all the bars are strained equally per square inch.

The following examples will illustrate the use of the tables:

A pin in the bolster or end shoe of a bridge has to carry a load of 40000 lbs. between two points of support; what size of pin is required, presuming the distance between points (*i. e.*, centers) of support of bolster plates and centers of pressure of end post plates $= 2\frac{1}{2}''$?

Answer: Bending moment $= 20000$ lbs. $\times 2\frac{1}{2} = 50000$ inch lbs., therefore $3\frac{1}{4}''$ pin required for 15000 lbs. fiber strain, since the allowed moment for $3\frac{1}{4}'' = 50600$, as per table.

Required the thickness of metal in the top chord or in a post of a bridge, that will give sufficient bearing area to a $3\frac{3}{8}''$ pin, having to transmit a stress of 63300 lbs., the allowed pressure per square inch on bearing being 12500 lbs. maximum.

The bearing value of a $3\frac{3}{8}''$ pin for $1''$ thickness of plate $= 42200$ lbs., therefore the thickness of metal required $= \frac{63300}{42200} = 1\frac{1}{2}''$, or each of the two plates in the chord or post will have to be $\frac{3}{4}''$ thick.

SHEARING AND BEARING VALUE OF RIVETS.

Bearing Value for different Thicknesses of Plate at 15000 lbs. per square inch.
(= Dia. of Rivet × Thickness of Plate × 15000 lbs.)

Diam. of Rivet in inches.		Area of Rivet.	Single Shear at 7500 lbs. per sq. inch.	$\frac{1}{4}''$	$\frac{5}{16}''$	$\frac{3}{8}''$	$\frac{7}{16}''$	$\frac{1}{2}''$	$\frac{9}{16}''$	$\frac{5}{8}''$	$\frac{11}{16}''$	$\frac{3}{4}''$	$\frac{13}{16}''$	$\frac{7}{8}''$
Fraction	Decimal.													
3/8	.375	.1104	828	1410										
7/16	.4375	.1503	1130	1640	2050									
1/2	.5	.1963	1470	1880	2340	2810								
9/16	.5625	.2485	1860	2110	2640	3160	3690							
5/8	.625	.3068	2300	2340	2930	3520	4100							
11/16	.6875	.3712	2780	2580	3220	3870	4510	5160						
3/4	.75	.4418	3310	2810	3520	4220	4920	5630	6330					
13/16	.8125	.5185	3890	3050	3810	4570	5330	6090	6860	7620				
7/8	.875	.6013	4510	3280	4100	4920	5740	6560	7380	8200				
15/16	.9375	.6903	5180	3520	4390	5270	6150	7030	7910	8790	9670			
1	1.0	.7854	5890	3750	4690	5620	6560	7500	8440	9380	10310	11250		
1 1/16	1.0625	.8866	6650	3980	4980	5980	6970	7970	8960	9960	10960	11950	12950	
1 1/8	1.125	.9940	7460	4220	5270	6330	7380	8440	9490	10550	11600	12660	13710	14770
1 3/16	1.1875	1.1075	8310	4450	5570	6680	7790	8910	10020	11130	12250	13360	14470	15590

MAXIMUM BENDING MOMENTS TO BE ALLOWED ON PINS FOR MAXIMUM FIBER STRAINS OF 15000, 20000 AND 22500 LBS. PER SQUARE INCH.

Diam. of Pin. Inches.	Moment for $S=15000$. Lbs. in.	Moment for $S=20000$. Lbs. in.	Moment for $S=22500$. Lbs. in.	Diam. of Pin. Inches.	Moment for $S=15000$. Lbs. in.	Moment for $S=20000$. Lbs. in.	Moment for $S=22500$. Lbs. in.
1	1470	1960	2210	4	94200	125700	141400
1 1/8	2100	2800	3140	4 1/8	103400	137800	155000
1 1/4	2880	3830	4310	4 1/4	113000	150700	169600
1 3/8	3830	5100	5740	4 3/8	123300	164400	185000
1 1/2	4970	6630	7460	4 1/2	134200	178900	201300
1 5/8	6320	8430	9480	4 5/8	145700	194300	218500
1 3/4	7890	10500	11800	4 3/4	157800	210400	236700
1 7/8	9710	12900	14600	4 7/8	170600	227500	255900
2	11800	15700	17700	5	184100	245400	276100
2 1/8	14100	18800	21200	5 1/8	198200	264300	297300
2 1/4	16800	22400	25200	5 1/4	213100	284100	319600
2 3/8	19700	26300	29600	5 3/8	228700	304900	343000
2 1/2	23000	30700	34500	5 1/2	245000	326700	367500
2 5/8	26600	35500	40000	5 5/8	262100	349500	393100
2 3/4	30600	40800	45900	5 3/4	280000	373300	419900
2 7/8	35000	46700	52500	5 7/8	298600	398200	447900
3	39800	53000	59600	6	318100	424100	477100
3 1/8	44900	59900	67400	6 1/8	338400	451200	507600
3 1/4	50600	67400	75800	6 1/4	359500	479400	539300
3 3/8	56600	75500	84900	6 3/8	381500	508700	572300
3 1/2	63100	84200	94700	6 1/2	404400	539200	606600
3 5/8	70100	93500	105200	6 5/8	428200	570900	642300
3 3/4	77700	103500	116500	6 3/4	452900	603900	679400
3 7/8	85700	114200	128500	6 7/8	478500	638000	717800

REMARKS—The following is the formula for flexure applied to pins:

$$M = \frac{S\pi d^3}{32} \quad \text{or} \quad = \frac{S A d}{8}$$

$M=$ moment of forces for any section of the pin.
$S=$ strain per sq. in. in extreme fibers of pin at that section.
$A=$ area of section.
$d=$ diameter.
$\pi = 3.14159$.

The forces are assumed to act in a plane passing through the axis of the pin.

The above table gives the values of M for different diameters of pin, and for three values of S.

If M max. is known, an inspection of the table will therefore show what diameter of pin must be used, in order that S does not exceed 15000, 20000 or 22500 lbs., as the requirements of the case may be.

For Railroad Bridges proportioned to a factor of safety of 5, it is customary to make S max. $= 15000$ lbs. in iron and $= 20000$ lbs. in steel.

BEARING VALUE OF PINS FOR ONE INCH THICKNESS OF PLATE.

(= Dia. of Pin × 1″ × strain per sq. inch.)

Diameter of Pin. Inches.	Area of Pin. Square Inches.	Bearing Value at 12500 lbs. per square inch. Lbs.	Bearing Value at 15000 lbs. per square inch. Lbs.
1	.785	12500	15000
1 1/8	.994	14100	16900
1 1/4	1.227	15600	18800
1 3/8	1.485	17200	20600
1 1/2	1.767	18800	22500
1 5/8	2.074	20300	24400
1 3/4	2.405	21900	26300
1 7/8	2.761	23400	28100
2	3.142	25000	30000
2 1/8	3.547	26600	31900
2 1/4	3.976	28100	33800
2 3/8	4.430	29700	35600
2 1/2	4.909	31300	37500
2 5/8	5.412	32800	39400
2 3/4	5.940	34400	41300
2 7/8	6.492	35900	43100
3	7.069	37500	45000
3 1/8	7.670	39100	46900
3 3/8	8.946	42200	50600
3 5/8	10.32	45300	54400
3 7/8	11.79	48400	58100
4 1/8	13.36	51600	61900
4 3/8	15.03	54700	65600
4 5/8	16.80	57800	69400
4 7/8	18.67	60900	73100
5 1/8	20.63	64100	76900
5 3/8	22.69	67200	80600
5 5/8	24.85	70300	84400
5 7/8	27.11	73400	88100
6 1/8	29.46	76600	91900
6 3/8	31.92	79700	95600
6 5/8	34.47	82800	99400
6 7/8	37.12	85900	103100

WOODEN BEAMS.

Safe Load, Uniformly Distributed, for Rectangular White or Yellow Pine Beams one inch thick,

allowing 1200 lbs. per square inch fiber strain.

To obtain the safe load for any thickness, multiply the safe load given in table, by the thickness of beam.

To obtain the required thickness for any load, divide by the safe load for 1 inch, given in table.

Span in Feet.	DEPTH OF BEAM.										
	6″	7″	8″	9″	10″	11″	12″	13″	14″	15″	16″
Feet.	Lbs.	Lbs.	Lbs.	Lbs.	Lbs.	Lbs.	Lbs.	Lbs.	Lbs.	Lbs.	Lbs.
5	960	1310	1710	2160	2670	3230	3840	4510	5230	6000	6830
6	800	1090	1420	1800	2220	2690	3200	3760	4360	5000	5690
7	690	930	1220	1540	1900	2300	2740	3220	3730	4290	4880
8	600	820	1070	1350	1670	2020	2400	2820	3270	3750	4270
9	530	730	950	1200	1480	1790	2130	2500	2900	3330	3790
10	480	650	850	1080	1330	1610	1920	2250	2610	3000	3410
11	440	590	780	980	1210	1470	1750	2050	2380	2730	3100
12	400	540	710	900	1110	1340	1600	1880	2180	2500	2840
13	370	500	660	830	1030	1240	1480	1730	2010	2310	2630
14	340	470	610	770	950	1150	1370	1610	1870	2140	2440
15	320	440	570	720	890	1080	1280	1500	1740	2000	2280
16	300	410	530	680	830	1010	1200	1410	1630	1880	2130
17	280	380	500	640	780	950	1130	1330	1540	1760	2010
18	270	360	470	600	740	900	1070	1250	1450	1670	1900
19	250	340	450	570	700	850	1010	1190	1380	1580	1800
20	240	330	430	540	670	810	960	1130	1310	1500	1710
21	230	310	410	510	630	770	910	1070	1240	1430	1630
22	220	300	390	490	610	730	870	1020	1190	1360	1550
23	210	280	370	470	580	700	830	980	1140	1300	1480
24	200	270	360	450	560	670	800	940	1090	1250	1420
25	190	260	340	430	530	650	770	900	1050	1200	1370
26	180	250	330	420	510	620	740	870	1010	1150	1310
27	180	240	320	400	500	600	710	830	970	1110	1260
28	170	230	300	390	480	580	690	800	930	1070	1220
29	170	230	290	370	460	560	660	780	900	1030	1180

EXPLANATION OF TABLES ON MAXIMUM STRESSES IN PRATT AND WHIPPLE TRUSSES.

Pages 141 to 143, inclusive.

These tables give the stress in each member of a Pratt (single quadrangular) or Whipple (double quadrangular) truss, for any number of panels not exceeding twelve in the former, and twenty in the latter case, on the assumption that the load is uniform per foot, and the panels are all of the same length. The stresses are given in terms of the truss-panel dead and moving loads, represented respectively by W and L. These are obtained by multiplying the dead load per foot of bridge, in the case of W, and the moving or live load per foot of bridge, in the case of L, by half the panel length.

The letters W and L are placed at the top of column, in tables, and not next to the figures to which they belong, for want of space.

The stress in aB, for example, in a twelve panel Pratt truss, $= 5.5 \text{ W} \times 5.5 \text{ L}$, and in Bc $= 4.5 \text{ W} \times \frac{5\frac{5}{8}}{1\frac{1}{2}} \text{ L}$, both multiplied by the quotient specified in the last column.

The system of lettering employed is shown by Figs. 7 and 8, on page 26 of the lithographs, and, it is believed, is the best in use. By making a sketch of the truss under consideration and lettering the vertices in the manner shown, the truss members to which reference is had in the tables, can be readily identified.

In the following tables, the dead load is assumed as concentrated at the lower vertices of the trusses, for through bridges, and at the upper vertices, for deck bridges. For through bridges of very large span, the stresses thus obtained for the posts must be increased by the truss-panel weight of the upper portion of the truss, including the lateral bracing; but in small spans, the increase of stress on this account is so inconsiderable that it is usually neglected.

Note: In order to calculate the stresses in a Whipple or double quadrangular truss by statical methods, it is necessary to consider the truss as the combination of two Pratt trusses or single systems of bracing, and assume that each of these two systems is strained in the same manner as if one were independent of the other. If the number of panels is odd, each of the two systems is unsymmetrical, which has the effect of making the stress in the middle panel of the lower chord slightly smaller than the stress in the corresponding panel of the top chord. To avoid this peculiarity and obtain equal stresses in these members, a division into symmetrical systems is sometimes assumed for the dead load stresses and for the full load, by considering the counter ties canceled. For the live load stresses obtained by partial loading, however, it is again necessary to divide into unsymmetrical systems, so that, while there appears to be no good reason in favor of this method, it has the objection of inconsistency. The difference in the resulting stresses obtained by the two methods is so small as not to be of practical consequence. Each of the two systems is assumed to carry one-half of the panel load at the top of the inclined end posts.

ILLUSTRATION OF APPLICATION OF TABLES, ALSO OF THE USE OF TABLE OF NATURAL SINES, TANGENTS AND SECANTS.

A Pratt truss of 135′ span and 18′ depth, is divided into nine panels of 15′ each. Required the stress in first main tie Bc, and in middle panel DE of top chord, for a dead load of 1200 lbs. and a moving load of 3000 lbs. per lineal foot of bridge.

$$W = \frac{1200}{2} \times 15 = 9000 \text{ lbs.}$$

$$L = \frac{3000}{2} \times 15 = 22500 \text{ lbs.}$$

$$Bc = (3\ W + \frac{28}{9} L) \times \frac{\text{Length Bc}}{18}$$

$$DE = (10\ W + 10\ L) \frac{15}{18}$$

The factor $\frac{15}{18}$, or panel length divided by depth of truss, is the tangent of the angle, for which the length Bc, divided by depth of truss, is the secant. By table of natural sines, tangents and secants, for tangent $= \frac{15}{18} = 0.833$, the secant $= 1.302$; therefore

$$Bc = 97000 \times 1.30 = 126100 \text{ lbs.}$$

$$DE = 315000 \times \frac{15}{18} = 262500 \text{ lbs.}$$

MAXIMUM STRESSES UNDER DEAD AND MOVING LOADS IN PRATT OR SINGLE QUADRANGULAR TRUSSES

With inclined end posts and equal panels, for Through and Deck Bridges.
W = dead load and L = moving load per truss and per panel.

Member.	12 Panel Truss.	11 Panel Truss.	10 Panel Truss.	9 Panel Truss.	8 Panel Truss.	Multiply by:
	$W+L$	$W+L$	$W+L$	$W+L$	$W+L$	
aB	$5.5+5.5$	$5+5$	$4.5+4.5$	$4+4$	$3.5+3.5$	Length of member divided by depth of truss.
Bc	$4.5+\frac{55}{12}$	$4+\frac{45}{11}$	$3.5+3.6$	$3+\frac{28}{9}$	$2.5+\frac{21}{8}$	
Cd	$3.5+\frac{45}{12}$	$3+\frac{36}{11}$	$2.5+2.8$	$2+\frac{21}{9}$	$1.5+\frac{15}{8}$	
De	$2.5+\frac{36}{12}$	$2+\frac{28}{11}$	$1.5+2.1$	$1+\frac{15}{9}$	$0.5+\frac{10}{8}$	
Ef	$1.5+\frac{28}{12}$	$1+\frac{21}{11}$	$0.5+1.5$	$0+\frac{10}{9}$	$-0.5+\frac{8}{8}$	
Fg	$0.5+\frac{21}{12}$	$0+\frac{15}{11}$	$-0.5+1.0$	$-1+\frac{6}{9}$	$-1.5+\frac{3}{8}$	
Gh	$-0.5+\frac{15}{12}$	$-1+\frac{10}{11}$	$-1.5+0.6$	$-2+\frac{3}{9}$		
Hi	$-1.5+\frac{10}{12}$	$-2+\frac{6}{11}$				
abc	$5.5+5.5$	$5+5$	$4.5+4.5$	$4+4$	$3.5+3.5$	Panel length divided by depth of truss.
BC, cd	$10.0+10.0$	$9+9$	$8.0+8.0$	$7+7$	$6.0+6.0$	
CD, de	$13.5+13.5$	$12+12$	$10.5+10.5$	$9+9$	$7.5+7.5$	
DE, ef	$16.0+16.0$	$14+14$	$12.0+12.0$	$10+10$	$8.0+8.0$	
EF, fg	$17.5+17.5$	$15+15$	$12.5+12.5$			
FG	$18.0+18.0$					
Thro'. Deck.						
Cc	$4.5+\frac{55}{12}$	$4+\frac{45}{11}$	$3.5+3.6$	$3+\frac{28}{9}$	$2.5+\frac{21}{8}$	Unity.
Cc, Dd	$3.5+\frac{45}{12}$	$3+\frac{36}{11}$	$2.5+2.8$	$2+\frac{21}{9}$	$1.5+\frac{15}{8}$	
Dd, Ee	$2.5+\frac{36}{12}$	$2+\frac{28}{11}$	$1.5+2.1$	$1+\frac{15}{9}$	$0.5+\frac{10}{8}$	
Ee, Ff	$1.5+\frac{28}{12}$	$1+\frac{21}{11}$	$0.5+1.5$	$0+\frac{10}{9}$	$-0.5+\frac{8}{8}$	
Ff, Gg	$0.5+\frac{21}{12}$	$0+\frac{15}{11}$	$-0.5+1.0$			
Gg	$-0.5+\frac{15}{12}$					

Member.	7 Panel Truss.	6 Panel Truss.	5 Panel Truss.	4 Panel Truss.	3 Panel Truss.	Multiply by:
	$W+L$	$W+L$	$W+L$	$W+L$	$W+L$	
aB	$3+3$	$2.5+2.5$	$2+2.0$	$1.5+1.5$	$1+1$	Length of member divided by depth of truss.
Bc	$2+\frac{15}{7}$	$1.5+\frac{10}{6}$	$1+1.2$	$0.5+\frac{3}{4}$	$0+\frac{1}{3}$	
Cd	$1+\frac{10}{7}$	$0.5+1.0$	$0+0.6$	$-0.5+\frac{1}{4}$		
De	$0+\frac{6}{7}$	$-0.5+0.5$	$-1+0.2$			
Ef	$-1+\frac{3}{7}$					
abc	$3+3$	$2.5+2.5$	$2+2$	$1.5+1.5$	$1+1$	Panel Length divided by depth of truss.
BC, cd	$5+5$	$4.0+4.0$	$3+3$	$2.0+2.0$	$1+1$	
CDE, de	$6+6$	$4.5+4.5$				
Thro'. Deck.						
Cc	$2+\frac{15}{7}$	$1.5+\frac{10}{6}$	$1+1.2$	$0.5+\frac{3}{4}$		Unity.
Cc, Dd	$1+\frac{10}{7}$	$0.5+1.0$	$0+0.6$	$-0.5+\frac{1}{4}$		
Dd	$0+\frac{6}{7}$	$-0.5+0.5$				

MAXIMUM STRESSES UNDER DEAD AND MOVING LOADS IN WHIPPLE OR DOUBLE QUADRANGULAR TRUSSES

With inclined end posts and equal panels, for Through and Deck Bridges.
W = dead load and L = moving load per truss and per panel.

Member.			20 Panel Truss. W+L	19 Panel Truss. W+L	18 Panel Truss. W+L	17 Panel Truss. W+L	16 Panel Truss. W+L	Multiply by:
		aB	$9.5+9.5$	$9+9$	$8.5+8.5$	$8+8$	$7.5+7.5$	
		Bc	$4.5+\frac{90.5}{20}$	$\frac{80}{19}+\frac{80.5}{19}$	$4.0+\frac{72.5}{18}$	$\frac{63}{17}+\frac{63.5}{17}$	$3.5+\frac{56.5}{16}$	
		Bd	$4.0+\frac{80.5}{20}$	$\frac{72}{19}+\frac{72.5}{19}$	$3.5+\frac{63.5}{18}$	$\frac{56}{17}+\frac{56.5}{17}$	$3.0+\frac{48.5}{16}$	
		Ce	$3.5+\frac{72.5}{20}$	$\frac{61}{19}+\frac{63.5}{19}$	$3.0+\frac{56.5}{18}$	$\frac{46}{17}+\frac{48.5}{17}$	$2.5+\frac{42.5}{16}$	Length of member divided by depth of truss.
		Df	$3.0+\frac{63.5}{20}$	$\frac{53}{19}+\frac{55.5}{19}$	$2.5+\frac{48.5}{18}$	$\frac{39}{17}+\frac{42.5}{17}$	$2.0+\frac{35.5}{16}$	
		Eg	$2.5+\frac{56.5}{20}$	$\frac{42}{19}+\frac{48.5}{19}$	$2.0+\frac{42.5}{18}$	$\frac{29}{17}+\frac{35.5}{17}$	$1.5+\frac{30.5}{16}$	
		Fh	$2.0+\frac{48.5}{20}$	$\frac{34}{19}+\frac{42.5}{19}$	$1.5+\frac{35.5}{18}$	$\frac{22}{17}+\frac{30.5}{17}$	$1.0+\frac{24.5}{16}$	
		Gi	$1.5+\frac{42.5}{20}$	$\frac{23}{19}+\frac{35.5}{19}$	$1.0+\frac{30.5}{18}$	$\frac{17}{17}+\frac{24.5}{17}$	$0.5+\frac{20.5}{16}$	
		Hk	$1.0+\frac{35.5}{20}$	$\frac{15}{19}+\frac{30.5}{19}$	$0.5+\frac{24.5}{18}$	$\frac{?}{17}+\frac{20.5}{17}$	$0.0+\frac{15.5}{16}$	
		Il	$0.5+\frac{30.5}{20}$	$\frac{4}{19}+\frac{24.5}{19}$	$0.0+\frac{20.5}{18}$	$-\frac{5}{17}+\frac{15.5}{17}$	$-0.5+\frac{12.5}{16}$	
		Km	$0.0+\frac{24.5}{20}$	$-\frac{5}{19}+\frac{20.5}{19}$	$-0.5+\frac{15.5}{18}$	$-\frac{12}{17}+\frac{12.5}{17}$	$-1.0+\frac{8.5}{16}$	
		Ln	$-0.5+\frac{20.5}{20}$	$-\frac{15}{19}+\frac{15.5}{19}$	$-1.0+\frac{12.5}{18}$	$-\frac{22}{17}+\frac{8.5}{17}$	$-1.5+\frac{6.5}{16}$	
		Mo	$-1.0+\frac{15.5}{20}$	$-\frac{23}{19}+\frac{12.5}{19}$				
		abc	$9.5+9.5$	$9+9$	$8.5+8.5$	$8+8$	$7.5+7.5$	
		cd	$14+14$	$\frac{251}{19}+\frac{251}{19}$	$12.5+12.5$	$\frac{199}{17}+\frac{199}{17}$	$11+11$	
BC,		de	$22+22$	$\frac{395}{19}+\frac{395}{19}$	$19.5+19.5$	$\frac{311}{17}+\frac{311}{17}$	$17+17$	
CD,		ef	$29+29$	$\frac{517}{19}+\frac{517}{19}$	$25.5+25.5$	$\frac{403}{17}+\frac{403}{17}$	$22+22$	
DE,		fg	$35+35$	$\frac{623}{19}+\frac{623}{19}$	$30.5+30.5$	$\frac{481}{17}+\frac{481}{17}$	$26+26$	Panel length divided by depth of truss.
EF,		gh	$40+40$	$\frac{707}{19}+\frac{707}{19}$	$34.5+34.5$	$\frac{539}{17}+\frac{539}{17}$	$29+29$	
FG,		hi	$44+44$	$\frac{775}{19}+\frac{775}{19}$	$37.5+37.5$	$\frac{583}{17}+\frac{583}{17}$	$31+31$	
GH,		ik	$47+47$	$\frac{821}{19}+\frac{821}{19}$	$39.5+39.5$	$\frac{607}{17}+\frac{607}{17}*$	$32+32$	
HI,		kl	$49+49$	$\frac{851}{19}+\frac{851}{19}*$	$40.5+40.5$	$\frac{617}{17}+\frac{617}{17}$	HI=GH	
IKL			$50+50$	$\frac{859}{19}+\frac{859}{19}$	IK=HI	IK=HI		
				*kl= $\frac{843}{19}+\frac{843}{19}$		*ik= $\frac{597}{17}+\frac{597}{17}$		
Thro'.	Deck.							
		Cc	$4.5+\frac{90.5}{20}$	$\frac{80}{19}+\frac{80.5}{19}$	$4.0+\frac{72.5}{18}$	$\frac{63}{17}+\frac{63.5}{17}$	$3.5+\frac{56.5}{16}$	
		Dd	$4.0+\frac{80.5}{20}$	$\frac{72}{19}+\frac{72.5}{19}$	$3.5+\frac{63.5}{18}$	$\frac{56}{17}+\frac{56.5}{17}$	$3.0+\frac{48.5}{16}$	
Cc,		Ee	$3.5+\frac{72.5}{20}$	$\frac{61}{19}+\frac{63.5}{19}$	$3.0+\frac{56.5}{18}$	$\frac{46}{17}+\frac{48.5}{17}$	$2.5+\frac{42.5}{16}$	
Dd,		Ff	$3.0+\frac{63.5}{20}$	$\frac{53}{19}+\frac{55.5}{19}$	$2.5+\frac{48.5}{18}$	$\frac{39}{17}+\frac{42.5}{17}$	$2.0+\frac{35.5}{16}$	
Ee,		Gg	$2.5+\frac{56.5}{20}$	$\frac{43}{19}+\frac{48.5}{19}$	$2.0+\frac{42.5}{18}$	$\frac{29}{17}+\frac{35.5}{17}$	$1.5+\frac{30.5}{16}$	
Ff,		Hh	$2.0+\frac{48.5}{20}$	$\frac{34}{19}+\frac{42.5}{19}$	$1.5+\frac{35.5}{18}$	$\frac{22}{17}+\frac{30.5}{17}$	$1.0+\frac{24.5}{16}$	Unity.
Gg,		Ii	$1.5+\frac{42.5}{20}$	$\frac{23}{19}+\frac{35.5}{19}$	$1.0+\frac{30.5}{18}$	$\frac{12}{17}+\frac{24.5}{17}$	$0.5+\frac{20.5}{16}$	
Hh,		Kk	$1.0+\frac{35.5}{20}$	$\frac{15}{19}+\frac{30.5}{19}$	$0.5+\frac{24.5}{18}$	$\frac{5}{17}+\frac{20.5}{17}$	$0.0+\frac{15.5}{16}$	
Ii,		Ll	$0.5+\frac{30.5}{20}$	$\frac{4}{19}+\frac{24.5}{19}$	$0.0+\frac{20.5}{18}$	$-\frac{5}{17}+\frac{15.5}{17}$	$-0.5+\frac{12.5}{16}$	
Kk			$0.0+\frac{24.5}{20}$	$-\frac{4}{19}+\frac{20.5}{19}$	$-0.5+\frac{15.5}{18}$			
Ll			$-0.5+\frac{20.5}{20}$					

MAXIMUM STRESSES UNDER DEAD AND MOVING LOADS IN WHIPPLE OR DOUBLE QUADRANGULAR TRUSSES

With inclined end posts and equal panels, for Through and Deck Bridges.

W = dead load and L = moving load per truss and per panel.

Member.	15 Panel Truss. $W+L$	14 Panel Truss. $W+L$	13 Panel Truss. $W+L$	12 Panel Truss. $W+L$	11 Panel Truss. $W+L$	Multiply by:
aB	$7+7$	$6.5+6.5$	$6+6$	$5.5+5.5$	$5+5$	Length of member divided by depth of truss.
Bc	$\frac{48}{15}+\frac{48.5}{15}$	$3.0+\frac{42.5}{14}$	$\frac{35}{13}+\frac{35.5}{13}$	$2.5+\frac{30.5}{12}$	$\frac{24}{11}+\frac{24.5}{11}$	
Bd	$\frac{42}{15}+\frac{42.5}{15}$	$2.5+\frac{35.5}{14}$	$\frac{30}{13}+\frac{30.5}{13}$	$2.0+\frac{24.5}{12}$	$\frac{20}{11}+\frac{20.5}{11}$	
Ce	$\frac{33}{15}+\frac{35.5}{15}$	$2.0+\frac{30.5}{14}$	$\frac{22}{13}+\frac{24.5}{13}$	$1.5+\frac{20.5}{12}$	$\frac{13}{11}+\frac{15.5}{11}$	
Df	$\frac{27}{15}+\frac{30.5}{15}$	$1.5+\frac{24.5}{14}$	$\frac{17}{13}+\frac{20.5}{13}$	$1.0+\frac{15.5}{12}$	$\frac{9}{11}+\frac{12.5}{11}$	
Eg	$\frac{18}{15}+\frac{24.5}{15}$	$1.0+\frac{20.5}{14}$	$\frac{9}{13}+\frac{15.5}{13}$	$0.5+\frac{12.5}{12}$	$\frac{2}{11}+\frac{8.5}{11}$	
Fh	$\frac{12}{15}+\frac{20.5}{15}$	$0.5+\frac{15.5}{14}$	$\frac{4}{13}+\frac{12.5}{13}$	$0.0+\frac{8.5}{12}$	$-\frac{2}{11}+\frac{6.5}{11}$	
Gi	$\frac{3}{15}+\frac{15.5}{15}$	$0.0+\frac{12.5}{14}$	$-\frac{4}{13}+\frac{8.5}{13}$	$-0.5+\frac{6.5}{12}$	$-\frac{9}{11}+\frac{3.5}{11}$	
Hk	$-\frac{3}{15}+\frac{12.5}{15}$	$-0.5+\frac{8.5}{14}$	$-\frac{9}{13}+\frac{6.5}{13}$	$-1.0+\frac{3.5}{12}$	$-\frac{13}{11}+\frac{2.5}{11}$	
Il	$-\frac{12}{15}+\frac{8.5}{15}$	$-1.0+\frac{6.5}{14}$	$-\frac{17}{13}+\frac{3.5}{13}$			
Km	$-\frac{18}{15}+\frac{6.5}{15}$					
abc	$7+7$	$6.5+6.5$	$6+6$	$5.5+5.5$	$5+5$	Panel length divided by depth of truss.
cd	$\frac{153}{15}+\frac{153}{15}$	$9.5+9.5$	$\frac{113}{13}+\frac{113}{13}$	$8.0+8.0$	$\frac{79}{11}+\frac{79}{11}$	
BC, de	$\frac{237}{15}+\frac{237}{15}$	$14.5+14.5$	$\frac{173}{13}+\frac{173}{13}$	$12.0+12.0$	$\frac{119}{11}+\frac{119}{11}$	
CD, ef	$\frac{303}{15}+\frac{303}{15}$	$18.5+18.5$	$\frac{217}{13}+\frac{217}{13}$	$15.0+15.0$	$\frac{145}{11}+\frac{145}{11}$	
DE, fg	$\frac{357}{15}+\frac{357}{15}$	$21.5+21.5$	$\frac{251}{13}+\frac{251}{13}$	$17.0+17.0$	$\frac{163}{11}+\frac{163}{11}*$	
EF, gh	$\frac{393}{15}+\frac{393}{15}$	$23.5+23.5$	$\frac{269}{13}+\frac{269}{13}*$	$18.0+18.0$	$\frac{167}{11}+\frac{167}{11}$	
FG, hi	$\frac{417}{15}+\frac{417}{15}*$	$24.5+24.5$	$\frac{277}{13}+\frac{277}{13}$	FG=EF	FG=EF	
GHI	$\frac{423}{15}+\frac{423}{15}$	GH=FG	GH=FG		*fg= $\frac{159}{11}+\frac{159}{11}$	
*hi= $\frac{411}{15}+\frac{411}{15}$		*gh= $\frac{261}{13}+\frac{261}{13}$				
Thro'. Deck.						Unity.
Cc	$\frac{48}{15}+\frac{48.5}{15}$	$3.0+\frac{42.5}{14}$	$\frac{35}{13}+\frac{35.5}{13}$	$2.5+\frac{30.5}{12}$	$\frac{24}{11}+\frac{24.5}{11}$	
Dd	$\frac{42}{15}+\frac{42.5}{15}$	$2.5+\frac{35.5}{14}$	$\frac{30}{13}+\frac{30.5}{13}$	$2.0+\frac{24.5}{12}$	$\frac{20}{11}+\frac{20.5}{11}$	
Cc, Ee	$\frac{33}{15}+\frac{35.5}{15}$	$2.0+\frac{30.5}{14}$	$\frac{22}{13}+\frac{24.5}{13}$	$1.5+\frac{20.5}{12}$	$\frac{13}{11}+\frac{15.5}{11}$	
Dd, Ff	$\frac{27}{15}+\frac{30.5}{15}$	$1.5+\frac{24.5}{14}$	$\frac{17}{13}+\frac{20.5}{13}$	$1.0+\frac{15.5}{12}$	$\frac{9}{11}+\frac{12.5}{11}$	
Ee, Gg	$\frac{18}{15}+\frac{24.5}{15}$	$1.0+\frac{20.5}{14}$	$\frac{9}{13}+\frac{15.5}{13}$	$0.5+\frac{12.5}{12}$	$\frac{2}{11}+\frac{8.5}{11}$	
Ff, Hh	$\frac{12}{15}+\frac{20.5}{15}$	$0.5+\frac{15.5}{14}$	$\frac{4}{13}+\frac{12.5}{13}$	$0.0+\frac{8.5}{12}$	$-\frac{2}{11}+\frac{6.5}{11}$	
Gg	$\frac{3}{15}+\frac{15.5}{15}$	$0.0+\frac{12.5}{14}$	$-\frac{4}{13}+\frac{8.5}{13}$	$-0.5+\frac{6.5}{12}$		
Hh	$-\frac{3}{15}+\frac{12.5}{15}$	$-0.5+\frac{8.5}{14}$				

NATURAL SINES, TANGENTS AND SECANTS,

Advancing by 10 min.

Deg.	Min.	Sine.	Tangent.	Secant.	Deg.	Min.	Sine.	Tangent.	Secant.
0	00	.0000	.0000	1.0000	5	00	.0872	.0875	1.0038
	10	.0029	.0029	1.0000		10	.0901	.0904	1.0041
	20	.0058	.0058	1.0000		20	.0929	.0934	1.0043
	30	.0087	.0087	1.0000		30	.0958	.0963	1.0046
	40	.0116	.0116	1.0001		40	.0987	.0992	1.0049
	50	.0145	.0145	1.0001		50	.1016	.1022	1.0052
1	00	.0175	.0175	1.0002	6	00	.1045	.1051	1.0055
	10	.0204	.0204	1.0002		10	.1074	.1080	1.0058
	20	.0233	.0233	1.0003		20	.1103	.1110	1.0061
	30	.0262	.0262	1.0003		30	.1132	.1139	1.0065
	40	.0291	.0291	1.0004		40	.1161	.1169	1.0068
	50	.0320	.0320	1.0005		50	.1190	.1198	1.0072
2	00	.0349	.0349	1.0006	7	00	.1219	.1228	1.0075
	10	.0378	.0378	1.0007		10	.1248	.1257	1.0079
	20	.0407	.0407	1.0008		20	.1276	.1287	1.0082
	30	.0436	.0437	1.0010		30	.1305	.1317	1.0086
	40	.0465	.0466	1.0011		40	.1334	.1346	1.0090
	50	.0494	.0495	1.0012		50	.1363	.1376	1.0094
3	00	.0523	.0524	1.0014	8	00	.1392	.1405	1.0098
	10	.0552	.0553	1.0015		10	.1421	.1435	1.0102
	20	.0581	.0582	1.0017		20	.1449	.1465	1.0107
	30	.0610	.0612	1.0019		30	.1478	.1495	1.0111
	40	.0640	.0641	1.0021		40	.1507	.1524	1.0116
	50	.0669	.0670	1.0022		50	.1536	.1554	1.0120
4	00	.0698	.0699	1.0024	9	00	.1564	.1584	1.0125
	10	.0727	.0729	1.0027		10	.1593	.1614	1.0129
	20	.0756	.0758	1.0029		20	.1622	.1644	1.0134
	30	.0785	.0787	1.0031		30	.1650	.1673	1.0139
	40	.0814	.0816	1.0033		40	.1679	.1703	1.0144
	50	.0843	.0846	1.0036		50	.1708	.1733	1.0149

NATURAL SINES, TANGENTS AND SECANTS.

(CONTINUED.)

Deg.	Min.	Sine.	Tangent.	Secant.	Deg.	Min.	Sine.	Tangent.	Secant.
10	00	.1736	.1763	1.0154	15	00	.2588	.2679	1.0353
	10	.1765	.1793	1.0160		10	.2616	.2711	1.0361
	20	.1794	.1823	1.0165		20	.2644	.2742	1.0369
	30	.1822	.1853	1.0170		30	.2672	.2773	1.0377
	40	.1851	.1883	1.0176		40	.2700	.2805	1.0386
	50	.1880	.1914	1.0181		50	.2728	.2836	1.0394
11	00	.1908	.1944	1.0187	16	00	.2756	.2867	1.0403
	10	.1937	.1974	1.0193		10	.2784	.2899	1.0412
	20	.1965	.2004	1.0199		20	.2812	.2931	1.0421
	30	.1994	.2035	1.0205		30	.2840	.2962	1.0429
	40	.2022	.2065	1.0211		40	.2868	.2994	1.0439
	50	.2051	.2095	1.0217		50	.2896	.3026	1.0448
12	00	.2079	.2126	1.0223	17	00	.2924	.3057	1.0457
	10	.2108	.2156	1.0230		10	.2952	.3089	1.0466
	20	.2136	.2186	1.0236		20	.2979	.3121	1.0476
	30	.2164	.2217	1.0243		30	.3007	.3153	1.0485
	40	.2193	.2247	1.0249		40	.3035	.3185	1.0495
	50	.2221	.2278	1.0256		50	.3062	.3217	1.0505
13	00	.2250	.2309	1.0263	18	00	.3090	.3249	1.0515
	10	.2278	.2339	1.0270		10	.3118	.3281	1.0525
	20	.2306	.2370	1.0277		20	.3145	.3314	1.0535
	30	.2334	.2401	1.0284		30	.3173	.3346	1.0545
	40	.2363	.2432	1.0291		40	.3201	.3378	1.0555
	50	.2391	.2462	1.0299		50	.3228	.3411	1.0566
14	00	.2419	.2493	1.0306	19	00	.3256	.3443	1.0576
	10	.2447	.2524	1.0314		10	.3283	.3476	1.0587
	20	.2476	.2555	1.0321		20	.3311	.3508	1.0598
	30	.2504	.2586	1.0329		30	.3338	.3541	1.0608
	40	.2532	.2617	1.0337		40	.3365	.3574	1.0619
	50	.2560	.2648	1.0345		50	.3393	.3607	1.0631

NATURAL SINES, TANGENTS AND SECANTS.

(CONTINUED.)

Deg.	Min.	Sine.	Tangent.	Secant.	Deg.	Min.	Sine.	Tangent.	Secant.
20	00	.3420	.3640	1.0642	25	00	.4226	.4663	1.1034
	10	.3448	.3673	1.0653		10	.4253	.4699	1.1049
	20	.3475	.3706	1.0665		20	.4279	.4734	1.1064
	30	.3502	.3739	1.0676		30	.4305	.4770	1.1079
	40	.3529	.3772	1.0688		40	.4331	.4806	1.1095
	50	.3557	.3805	1.0700		50	.4358	.4841	1.1110
21	00	.3584	.3839	1.0711	26	00	.4384	.4877	1.1126
	10	.3611	.3872	1.0723		10	.4410	.4913	1.1142
	20	.3638	.3906	1.0736		20	.4436	.4950	1.1158
	30	.3665	.3939	1.0748		30	.4462	.4986	1.1174
	40	.3692	.3973	1.0760		40	.4488	.5022	1.1190
	50	.3719	.4006	1.0773		50	.4514	.5059	1.1207
22	00	.3746	.4040	1.0785	27	00	.4540	.5095	1.1223
	10	.3773	.4074	1.0798		10	.4566	.5132	1.1240
	20	.3800	.4108	1.0811		20	.4592	.5169	1.1257
	30	.3827	.4142	1.0824		30	.4617	.5206	1.1274
	40	.3854	.4176	1.0837		40	.4643	.5243	1.1291
	50	.3881	.4210	1.0850		50	.4669	.5280	1.1308
23	00	.3907	.4245	1.0864	28	00	.4695	.5317	1.1326
	10	.3934	.4279	1.0877		10	.4720	.5354	1.1343
	20	.3961	.4314	1.0891		20	.4746	.5392	1.1361
	30	.3987	.4348	1.0904		30	.4772	.5430	1.1379
	40	.4014	.4383	1.0918		40	.4797	.5467	1.1397
	50	.4041	.4417	1.0932		50	.4823	.5505	1.1415
24	00	.4067	.4452	1.0946	29	00	.4848	.5543	1.1434
	10	.4094	.4487	1.0961		10	.4874	.5581	1.1452
	20	.4120	.4522	1.0975		20	.4899	.5619	1.1471
	30	.4147	.4557	1.0989		30	.4924	.5658	1.1490
	40	.4173	.4592	1.1004		40	.4950	.5696	1.1509
	50	.4200	.4628	1.1019		50	.4975	.5735	1.1528

NATURAL SINES, TANGENTS AND SECANTS.

(CONTINUED.)

Deg.	Min.	Sine.	Tangent.	Secant.	Deg.	Min.	Sine.	Tangent.	Secant.
30	00	.5000	.5774	1.1547	35	00	.5736	.7002	1.2208
	10	.5025	.5812	1.1566		10	.5760	.7046	1.2233
	20	.5050	.5851	1.1586		20	.5783	.7089	1.2258
	30	.5075	.5890	1.1606		30	.5807	.7133	1.2283
	40	.5100	.5930	1.1626		40	.5831	.7177	1.2309
	50	.5125	.5969	1.1646		50	.5854	.7221	1.2335
31	00	.5150	.6009	1.1666	36	00	.5878	.7265	1.2361
	10	.5175	.6048	1.1687		10	.5901	.7310	1.2387
	20	.5200	.6088	1.1707		20	.5925	.7355	1.2413
	30	.5225	.6128	1.1728		30	.5948	.7400	1.2440
	40	.5250	.6168	1.1749		40	.5972	.7445	1.2467
	50	.5275	.6208	1.1770		50	.5995	.7490	1.2494
32	00	.5299	.6249	1.1792	37	00	.6018	.7536	1.2521
	10	.5324	.6289	1.1813		10	.6041	.7581	1.2549
	20	.5348	.6330	1.1835		20	.6065	.7627	1.2577
	30	.5373	.6371	1.1857		30	.6088	.7673	1.2605
	40	.5398	.6412	1.1879		40	.6111	.7720	1.2633
	50	.5422	.6453	1.1901		50	.6134	.7766	1.2661
33	00	.5446	.6494	1.1924	38	00	.6157	.7813	1.2690
	10	.5471	.6536	1.1946		10	.6180	.7860	1.2719
	20	.5495	.6577	1.1969		20	.6202	.7907	1.2748
	30	.5519	.6619	1.1992		30	.6225	.7954	1.2778
	40	.5544	.6661	1.2015		40	.6248	.8002	1.2808
	50	.5568	.6703	1.2039		50	.6271	.8050	1.2837
34	00	.5592	.6745	1.2062	39	00	.6293	.8098	1.2868
	10	.5616	.6787	1.2086		10	.6316	.8146	1.2898
	20	.5640	.6830	1.2110		20	.6338	.8195	1.2929
	30	.5664	.6873	1.2134		30	.6361	.8243	1.2960
	40	.5688	.6916	1.2158		40	.6383	.8292	1.2991
	50	.5712	.6959	1.2183		50	.6406	.8342	1.3022

NATURAL SINES, TANGENTS AND SECANTS.

(CONTINUED.)

Deg.	Min.	Sine.	Tangent.	Secant.	Deg.	Min.	Sine.	Tangent.	Secant.
40	00	.6428	.8391	1.3054	45	00	.7071	1.0000	1.4142
	10	.6450	.8441	1.3086		10	.7092	1.0058	1.4183
	20	.6472	.8491	1.3118		20	.7112	1.0117	1.4225
	30	.6494	.8541	1.3151		30	.7133	1.0176	1.4267
	40	.6517	.8591	1.3184		40	.7153	1.0235	1.4310
	50	.6539	.8642	1.3217		50	.7173	1.0295	1.4352
41	00	.6561	.8693	1.3250	46	00	.7193	1.0355	1.4396
	10	.6583	.8744	1.3284		10	.7214	1.0416	1.4439
	20	.6604	.8796	1.3318		20	.7234	1.0477	1.4483
	30	.6626	.8847	1.3352		30	.7254	1.0538	1.4527
	40	.6648	.8899	1.3386		40	.7274	1.0599	1.4572
	50	.6670	.8952	1.3421		50	.7294	1.0661	1.4617
42	00	.6691	.9004	1.3456	47	00	.7314	1.0724	1.4663
	10	.6713	.9057	1.3492		10	.7333	1.0786	1.4709
	20	.6734	.9110	1.3527		20	.7353	1.0850	1.4755
	30	.6756	.9163	1.3563		30	.7373	1.0913	1.4802
	40	.6777	.9217	1.3600		40	.7392	1.0977	1.4849
	50	.6799	.9271	1.3636		50	.7412	1.1041	1.4897
43	00	.6820	.9325	1.3673	48	00	.7431	1.1106	1.4945
	10	.6841	.9380	1.3711		10	.7451	1.1171	1.4993
	20	.6862	.9435	1.3748		20	.7470	1.1237	1.5042
	30	.6884	.9490	1.3786		30	.7490	1.1303	1.5092
	40	.6905	.9545	1.3824		40	.7509	1.1369	1.5141
	50	.6926	.9601	1.3863		50	.7528	1.1436	1.5192
44	00	.6947	.9657	1.3902	49	00	.7547	1.1504	1.5243
	10	.6967	.9713	1.3941		10	.7566	1.1571	1.5294
	20	.6988	.9770	1.3980		20	.7585	1.1640	1.5345
	30	.7009	.9827	1.4020		30	.7604	1.1708	1.5398
	40	.7030	.9884	1.4061		40	.7623	1.1778	1.5450
	50	.7050	.9942	1.4101		50	.7642	1.1847	1.5504

NATURAL SINES, TANGENTS AND SECANTS.

(CONTINUED.)

Deg.	Min.	Sine.	Tangent.	Secant.	Deg.	Min.	Sine.	Tangent.	Secant.
50	00	.7660	1.1918	1.5557	55	00	.8192	1.4281	1.7434
	10	.7679	1.1988	1.5611		10	.8208	1.4370	1.7507
	20	.7698	1.2059	1.5666		20	.8225	1.4460	1.7581
	30	.7716	1.2131	1.5721		30	.8241	1.4550	1.7655
	40	.7735	1.2203	1.5777		40	.8258	1.4641	1.7730
	50	.7753	1.2276	1.5833		50	.8274	1.4733	1.7806
51	00	.7771	1.2349	1.5890	56	00	.8290	1.4826	1.7883
	10	.7790	1.2423	1.5948		10	.8307	1.4919	1.7960
	20	.7808	1.2497	1.6005		20	.8323	1.5013	1.8039
	30	.7826	1.2572	1.6064		30	.8339	1.5108	1.8118
	40	.7844	1.2647	1.6123		40	.8355	1.5204	1.8198
	50	.7862	1.2723	1.6183		50	.8371	1.5301	1.8279
52	00	.7880	1.2799	1.6243	57	00	.8387	1.5399	1.8361
	10	.7898	1.2876	1.6303		10	.8403	1.5497	1.8443
	20	.7916	1.2954	1.6365		20	.8418	1.5597	1.8527
	30	.7934	1.3032	1.6427		30	.8434	1.5697	1.8612
	40	.7951	1.3111	1.6489		40	.8450	1.5798	1.8699
	50	.7969	1.3190	1.6553		50	.8465	1.5900	1.8783
53	00	.7986	1.3270	1.6616	58	00	.8480	1.6003	1.8871
	10	.8004	1.3351	1.6681		10	.8496	1.6107	1.8959
	20	.8021	1.3432	1.6746		20	.8511	1.6213	1.9048
	30	.8039	1.3514	1.6812		30	.8526	1.6319	1.9139
	40	.8056	1.3597	1.6878		40	.8542	1.6426	1.9230
	50	.8073	1.3680	1.6945		50	.8557	1.6534	1.9323
54	00	.8090	1.3764	1.7013	59	00	.8572	1.6643	1.9416
	10	.8107	1.3848	1.7081		10	.8587	1.6753	1.9511
	20	.8124	1.3934	1.7151		20	.8601	1.6864	1.9606
	30	.8141	1.4019	1.7221		30	.8616	1.6977	1.9703
	40	.8158	1.4106	1.7291		40	.8631	1.7090	1.9801
	50	.8175	1.4193	1.7362		50	.8646	1.7205	1.9900

NATURAL SINES, TANGENTS AND SECANTS.

(CONTINUED.)

Deg.	Min.	Sine.	Tangent.	Secant.	Deg.	Min.	Sine.	Tangent.	Secant.
60	00	.8660	1.7321	2.0000	65	00	.9063	2.1445	2.3662
	10	.8675	1.7437	2.0101		10	.9075	2.1609	2.3811
	20	.8689	1.7556	2.0204		20	.9088	2.1775	2.3961
	30	.8704	1.7675	2.0308		30	.9100	2.1943	2.4114
	40	.8718	1.7796	2.0413		40	.9112	2.2113	2.4269
	50	.8732	1.7917	2.0519		50	.9124	2.2286	2.4426
61	00	.8746	1.8040	2.0627	66	00	.9135	2.2460	2.4586
	10	.8760	1.8165	2.0736		10	.9147	2.2637	2.4748
	20	.8774	1.8291	2.0846		20	.9159	2.2817	2.4912
	30	.8788	1.8418	2.0957		30	.9171	2.2998	2.5078
	40	.8802	1.8546	2.1070		40	.9182	2.3183	2.5247
	50	.8816	1.8676	2.1185		50	.9194	2.3369	2.5419
62	00	.8829	1.8807	2.1301	67	00	.9205	2.3559	2.5593
	10	.8843	1.8940	2.1418		10	.9216	2.3750	2.5770
	20	.8857	1.9074	2.1537		20	.9228	2.3945	2.5949
	30	.8870	1.9210	2.1657		30	.9239	2.4141	2.6131
	40	.8884	1.9347	2.1786		40	.9250	2.4342	2.6316
	50	.8897	1.9486	2.1902		50	.9261	2.4545	2.6504
63	00	.8910	1.9626	2.2027	68	00	.9272	2.4751	2.6695
	10	.8923	1.9768	2.2153		10	.9283	2.4960	2.6888
	20	.8936	1.9912	2.2282		20	.9293	2.5172	2.7085
	30	.8949	2.0057	2.2412		30	.9304	2.5386	2.7285
	40	.8962	2.0204	2.2543		40	.9315	2.5605	2.7488
	50	.8975	2.0353	2.2677		50	.9325	2.5826	2.7695
64	00	.8988	2.0503	2.2812	69	00	.9336	2.6051	2.7904
	10	.9001	2.0655	2.2949		10	.9346	2.6279	2.8117
	20	.9013	2.0809	2.3088		20	.9356	2.6511	2.8334
	30	.9026	2.0965	2.3228		30	.9367	2.6746	2.8555
	40	.9038	2.1123	2.3371		40	.9377	2.6985	2.8779
	50	.9051	2.1283	2.3515		50	.9387	2.7228	2.9006

NATURAL SINES, TANGENTS AND SECANTS.

(CONTINUED.)

Deg.	Min.	Sine.	Tangent.	Secant.	Deg.	Min.	Sine.	Tangent.	Secant.
70	00	.9397	2.7475	2.9238	75	00	.9659	3.7321	3.8637
	10	.9407	2.7725	2.9474		10	.9667	3.7760	3.9061
	20	.9417	2.7980	2.9713		20	.9674	3.8208	3.9495
	30	.9426	2.8239	2.9957		30	.9681	3.8667	3.9939
	40	.9436	2.8502	3.0206		40	.9689	3.9136	4.0394
	50	.9446	2.8770	3.0458		50	.9696	3.9617	4.0859
71	00	.9455	2.9042	3.0716	76	00	.9703	4.0108	4.1336
	10	.9465	2.9319	3.0977		10	.9710	4.0611	4.1824
	20	.9474	2.9600	3.1244		20	.9717	4.1126	4.2324
	30	.9483	2.9887	3.1515		30	.9724	4.1653	4.2837
	40	.9492	3.0178	3.1792		40	.9730	4.2193	4.3362
	50	.9502	3.0475	3.2074		50	.9737	4.2747	4.3901
72	00	.9511	3.0777	3.2361	77	00	.9744	4.3315	4.4454
	10	.9520	3.1084	3.2653		10	.9750	4.3897	4.5022
	20	.9528	3.1397	3.2951		20	.9757	4.4494	4.5604
	30	.9537	3.1716	3.3255		30	.9763	4.5107	4.6202
	40	.9546	3.2041	3.3565		40	.9769	4.5736	4.6817
	50	.9555	3.2371	3.3881		50	.9775	4.6382	4.7448
73	00	.9563	3.2709	3.4203	78	00	.9781	4.7046	4.8097
	10	.9572	3.3052	3.4532		10	.9787	4.7729	4.8765
	20	.9580	3.3402	3.4867		20	.9793	4.8430	4.9452
	30	.9588	3.3759	3.5209		30	.9799	4.9152	5.0159
	40	.9596	3.4124	3.5559		40	.9805	4.9894	5.0886
	50	.9605	3.4495	3.5915		50	.9811	5.0658	5.1636
74	00	.9613	3.4874	3.6280	79	00	.9816	5.1446	5.2408
	10	.9621	3.5261	3.6652		10	.9822	5.2257	5.3205
	20	.9628	3.5656	3.7032		20	.9827	5.3093	5.4026
	30	.9636	3.6059	3.7420		30	.9833	5.3955	5.4874
	40	.9644	3.6470	3.7817		40	.9838	5.4845	5.5749
	50	.9652	3.6891	3.8222		50	.9843	5.5764	5.6653

NATURAL SINES, TANGENTS AND SECANTS.

(CONTINUED.)

Deg.	Min.	Sine.	Tangent.	Secant.	Deg.	Min.	Sine.	Tangent.	Secant.
80	00	.9848	5.6713	5.7588	85	00	.9962	11.430	11.474
	10	.9853	5.7694	5.8554		10	.9964	11.826	11.868
	20	.9858	5.8708	5.9554		20	.9967	12.251	12.291
	30	.9863	5.9758	6.0589		30	.9969	12.706	12.745
	40	.9868	6.0844	6.1661		40	.9971	13.197	13.235
	50	.9872	6.1970	6.2772		50	.9974	13.727	13.763
81	00	.9877	6.3138	6.3925	86	00	.9976	14.301	14.336
	10	.9881	6.4348	6.5121		10	.9978	14.924	14.958
	20	.9886	6.5606	6.6363		20	.9980	15.605	15.637
	30	.9890	6.6912	6.7655		30	.9981	16.350	16.380
	40	.9894	6.8269	6.8998		40	.9983	17.169	17.198
	50	.9899	6.9682	7.0396		50	.9985	18.075	18.103
82	00	.9903	7.1154	7.1853	87	00	.9986	19.081	19.107
	10	.9907	7.2687	7.3372		10	.9988	20.206	20.230
	20	.9911	7.4287	7.4957		20	.9989	21.470	21.494
	30	.9914	7.5958	7.6613		30	.9990	22.904	22.926
	40	.9918	7.7704	7.8344		40	.9992	24.542	24.562
	50	.9922	7.9530	8.0156		50	.9993	26.432	26.451
83	00	.9925	8.1443	8.2055	88	00	.9994	28.636	28.654
	10	.9929	8.3450	8.4047		10	.9995	31.242	31.258
	20	.9932	8.5555	8.6138		20	.9996	34.368	34.382
	30	.9936	8.7769	8.8337		30	.9997	38.188	38.202
	40	.9939	9.0098	9.0652		40	.9997	42.964	42.976
	50	.9942	9.2553	9.3092		50	.9998	49.104	49.114
84	00	.9945	9.5144	9.5668	89	00	.9998	57.290	57.299
	10	.9948	9.7882	9.8391		10	.9999	68.750	68.757
	20	.9951	10.0780	10.1275		20	.9999	85.940	85.946
	30	.9954	10.3854	10.4334		30	1.0000	114.589	114.593
	40	.9957	10.7119	10.7585		40	1.0000	171.885	171.888
	50	.9959	11.0594	11.1045		50	1.0000	343.774	343.775
					90	00	1.0000	Infinite.	Infinite.

LOGARITHMS OF NUMBERS.

No.	0	1	2	3	4	5	6	7	8	9	Diff.
10	0000	0043	0086	0128	0170	0212	0253	0294	0334	0374	40
11	0414	0453	0492	0531	0569	0607	0645	0682	0719	0755	37
12	0792	0828	0864	0899	0934	0969	1004	1038	1072	1106	33
13	1139	1173	1206	1239	1271	1303	1335	1367	1399	1430	31
14	1461	1492	1523	1553	1584	1614	1644	1673	1703	1732	29
15	1761	1790	1818	1847	1875	1903	1931	1959	1987	2014	27
16	2041	2068	2095	2122	2148	2175	2201	2227	2253	2279	25
17	2304	2330	2355	2380	2405	2430	2455	2480	2504	2529	24
18	2553	2577	2601	2625	2648	2672	2695	2718	2742	2765	23
19	2788	2810	2833	2856	2878	2900	2923	2945	2967	2989	21
20	3010	3032	3054	3075	3096	3118	3139	3160	3181	3201	21
21	3222	3243	3263	3284	3304	3324	3345	3365	3385	3404	20
22	3424	3444	3464	3483	3502	3522	3541	3560	3579	3598	19
23	3617	3636	3655	3674	3692	3711	3729	3747	3766	3784	18
24	3802	3820	3838	3856	3874	3892	3909	3927	3945	3962	17
25	3979	3997	4014	4031	4048	4065	4082	4099	4116	4133	17
26	4150	4166	4183	4200	4216	4232	4249	4265	4281	4298	16
27	4314	4330	4346	4362	4378	4393	4409	4425	4440	4456	16
28	4472	4487	4502	4518	4533	4548	4564	4579	4594	4609	15
29	4624	4639	4654	4669	4683	4698	4713	4728	4742	4757	14
30	4771	4786	4800	4814	4829	4843	4857	4871	4886	4900	14
31	4914	4928	4942	4955	4969	4983	4997	5011	5024	5038	13
32	5051	5065	5079	5092	5105	5119	5132	5145	5159	5172	13
33	5185	5198	5211	5224	5237	5250	5263	5276	5289	5302	13
34	5315	5328	5340	5353	5366	5378	5391	5403	5416	5428	13
35	5441	5453	5465	5478	5490	5502	5514	5527	5539	5551	12
36	5563	5575	5587	5599	5611	5623	5635	5647	5658	5670	12
37	5682	5694	5705	5717	5729	5740	5752	5763	5775	5786	12
38	5798	5809	5821	5832	5843	5855	5866	5877	5888	5899	12
39	5911	5922	5933	5944	5955	5966	5977	5988	5999	6010	11
No.	0	1	2	3	4	5	6	7	8	9	Diff.

LOGARITHMS OF NUMBERS—Continued.

No.	0	1	2	3	4	5	6	7	8	9	Diff.
40	6021	6031	6042	6053	6064	6075	6085	6096	6107	6117	11
41	6128	6138	6149	6160	6170	6180	6191	6201	6212	6222	10
42	6232	6243	6253	6263	6274	6284	6294	6304	6314	6325	10
43	6335	6345	6355	6365	6375	6385	6395	6405	6415	6425	10
44	6435	6444	6454	6464	6474	6484	6493	6503	6513	6522	10
45	6532	6542	6551	6561	6571	6580	6590	6599	6609	6618	10
46	6628	6637	6646	6656	6665	6675	6684	6693	6702	6712	9
47	6721	6730	6739	6749	6758	6767	6776	6785	6794	6803	9
48	6812	6821	6830	6839	6848	6857	6866	6875	6884	6893	9
49	6902	6911	6920	6928	6937	6946	6955	6964	6972	6981	9
50	6990	6998	7007	7016	7024	7033	7042	7050	7059	7067	9
51	7076	7084	7093	7101	7110	7118	7126	7135	7143	7152	8
52	7160	7168	7177	7185	7193	7202	7210	7218	7226	7235	8
53	7243	7251	7259	7267	7275	7284	7292	7300	7308	7316	8
54	7324	7332	7340	7348	7356	7364	7372	7380	7388	7396	8
55	7404	7412	7419	7427	7435	7443	7451	7459	7466	7474	8
56	7482	7490	7497	7505	7513	7520	7528	7536	7543	7551	8
57	7559	7566	7574	7582	7589	7597	7604	7612	7619	7627	7
58	7634	7642	7649	7657	7664	7672	7679	7686	7694	7701	8
59	7709	7716	7723	7731	7738	7745	7752	7760	7767	7774	8
60	7782	7789	7796	7803	7810	7818	7825	7832	7839	7846	7
61	7853	7860	7868	7875	7882	7889	7896	7903	7910	7917	7
62	7924	7931	7938	7945	7952	7959	7966	7973	7980	7987	6
63	7993	8000	8007	8014	8021	8028	8035	8041	8048	8055	7
64	8062	8069	8075	8082	8089	8096	8102	8109	8116	8122	7
65	8129	8136	8142	8149	8156	8162	8169	8176	8182	8189	6
66	8195	8202	8209	8215	8222	8228	8235	8241	8248	8254	7
67	8261	8267	8274	8280	8287	8293	8299	8306	8312	8319	6
68	8325	8331	8338	8344	8351	8357	8363	8370	8376	8382	6
69	8388	8395	8401	8407	8414	8420	8426	8432	8439	8445	6
No.	0	1	2	3	4	5	6	7	8	9	Diff.

LOGARITHMS OF NUMBERS—Continued.

No.	0	1	2	3	4	5	6	7	8	9	Diff.
70	8451	8457	8463	8470	8476	8482	8488	8494	8500	8506	7
71	8513	8519	8525	8531	8537	8543	8549	8555	8561	8567	6
72	8573	8579	8585	8591	8597	8603	8609	8615	8621	8627	6
73	8633	8639	8645	8651	8657	8663	8669	8675	8681	8686	6
74	8692	8698	8704	8710	8716	8722	8727	8733	8739	8745	6
75	8751	8756	8762	8768	8774	8779	8785	8791	8797	8802	6
76	8808	8814	8820	8825	8831	8837	8842	8848	8854	8859	6
77	8865	8871	8876	8882	8887	8893	8899	8904	8910	8915	6
78	8921	8927	8932	8938	8943	8949	8954	8960	8965	8971	5
79	8976	8982	8987	8993	8998	9004	9009	9015	9020	9025	6
80	9031	9036	9042	9047	9053	9058	9063	9069	9074	9079	6
81	9085	9090	9096	9101	9106	9112	9117	9122	9128	9133	5
82	9138	9143	9149	9154	9159	9165	9170	9175	9180	9186	5
83	9191	9196	9201	9206	9212	9217	9222	9227	9232	9238	5
84	9243	9248	9253	9258	9263	9269	9274	9279	9284	9289	5
85	9294	9299	9304	9309	9315	9320	9325	9330	9335	9340	5
86	9345	9350	9355	9360	9365	9370	9375	9380	9385	9390	5
87	9395	9400	9405	9410	9415	9420	9425	9430	9435	9440	5
88	9445	9450	9455	9460	9465	9469	9474	9479	9484	9489	5
89	9494	9499	9504	9509	9513	9518	9523	9528	9533	9538	4
90	9542	9547	9552	9557	9562	9566	9571	9576	9581	9586	4
91	9590	9595	9600	9605	9609	9614	9619	9624	9628	9633	5
92	9638	9643	9647	9652	9657	9661	9666	9671	9675	9680	5
93	9685	9689	9694	9699	9703	9708	9713	9717	9722	9727	4
94	9731	9736	9741	9745	9750	9754	9759	9763	9768	9773	4
95	9777	9782	9786	9791	9795	9800	9805	9809	9814	9818	5
96	9823	9827	9832	9836	9841	9845	9850	9854	9859	9863	5
97	9868	9872	9877	9881	9886	9890	9894	9899	9903	9908	4
98	9912	9917	9921	9926	9930	9934	9939	9943	9948	9952	4
99	9956	9961	9965	9969	9974	9978	9983	9987	9991	9996	4
No.	0	1	2	3	4	5	6	7	8	9	Diff.

WEIGHT OF
A CUBIC FOOT OF SUBSTANCES.

Names of Substances.	Average Weight Lbs.
Anthracite, solid, of Pennsylvania,	93
" broken, loose,	54
" " moderately shaken,	58
" heaped bushel, loose,	(80)
Ash, American white, dry,	38
Asphaltum,	87
Brass, (Copper and Zinc,) cast,	504
" rolled,	524
Brick, best pressed,	150
" common hard,	125
" soft, inferior,	100
Brickwork, pressed brick,	140
" ordinary,	112
Cement, hydraulic, ground, loose, American, Rosendale,	56
" " " " " Louisville,	50
" " " " English, Portland,	90
Cherry, dry,	42
Chestnut, dry,	41
Coal, bituminous, solid,	84
" " broken, loose,	49
" " heaped bushel, loose,	(74)
Coke, loose, of good coal,	27
" " heaped bushel,	(38)
Copper, cast,	542
" rolled,	548
Earth, common loam, dry, loose,	76
" " " " moderately rammed,	95
" as a soft flowing mud,	108
Ebony, dry,	76
Elm, dry,	35
Flint,	162
Glass, common window,	157

WEIGHT OF SUBSTANCES—Continued.

Names of Substances.	Average Weight. Lbs.
Gneiss, common,	168
Gold, cast, pure, or 24 carat,	1204
" pure, hammered,	1217
Granite,	170
Gravel, about the same as sand, which see.	
Hemlock, dry,	25
Hickory, dry,	53
Hornblende, black,	203
Ice,	58.7
Iron, cast,	450
" wrought, purest,	485
" " average,	480
Ivory,	114
Lead,	711
Lignum Vitæ, dry,	83
Lime, quick, ground, loose, or in small lumps,	53
" " " " thoroughly shaken,	75
" " " " per struck bushel,	(66)
Limestones and Marbles,	168
" " loose, in irregular fragments,	96
Mahogany, Spanish, dry,	53
" Honduras, dry,	35
Maple, dry,	49
Marbles, see Limestones.	
Masonry, of granite or limestone, well dressed,	165
" " mortar rubble,	154
" " dry " (well scabbled,)	138
" " sandstone, well dressed,	144
Mercury, at 32° Fahrenheit,	849
Mica,	183
Mortar, hardened,	103
Mud, dry, close,	80 to 110
" wet, fluid, maximum,	120
Oak, live, dry,	59

WEIGHT OF SUBSTANCES—Continued.

Names of Substances.	Average Weight. Lbs.
Oak, white, dry,	52
" other kinds,	32 to 45
Petroleum,	55
Pine, white, dry,	25
" yellow, Northern,	34
" " Southern,	45
Platinum,	1342
Quartz, common, pure,	165
Rosin,	69
Salt, coarse, Syracuse, N. Y.	45
" Liverpool, fine, for table use,	49
Sand, of pure quartz, dry, loose,	90 to 106
" well shaken,	99 to 117
" perfectly wet,	120 to 140
Sandstones, fit for building,	151
Shales, red or black,	162
Silver,	655
Slate,	175
Snow, freshly fallen,	5 to 12
" moistened and compacted by rain,	15 to 50
Spruce, dry,	25
Steel,	490
Sulphur,	125
Sycamore, dry,	37
Tar,	62
Tin, cast,	459
Turf or Peat, dry, unpressed,	20 to 30
Walnut, black, dry,	38
Water, pure rain or distilled, at 60° Fahrenheit,	$62\frac{1}{3}$
" sea,	64
Wax, bees,	60.5
Zinc or Spelter,	437

Green timbers usually weigh from one-fifth to one-half more than dry.

WINDOW GLASS.

Window Glass is sold by the box, which contains, as nearly as may be, 50 square feet, whatever may be the size of panes.

The thickness of ordinary or "single thick" Window Glass is about 1-16 of an inch, and of "double thick" nearly 1-8 of an inch.

The tensile strength of common glass varies from 2000 lbs. to 3000 lbs. per square inch, and its crushing strength from 6000 lbs. to 10000 lbs.

The following is the list of the Pittsburgh City Glass Works, Cunninghams & Co., Proprietors. Other sizes may be made to order.

Sizes. In.	Lights per Box. No.	Sizes. In.	Lights per Box. No.	Sizes. In.	Lights per Box. No.	Sizes. In.	Lights per Box. No.	Sizes. In.	Lights per Box. No.
6 × 8	150	11 × 24	27	14 × 18	29	18 × 18	22	24 × 30	10
7 × 9	115	26	25	20	26	20	20	32	10
8 × 10	90	28	23	22	24	22	18	34	9
12	75	30	22	24	22	24	17	36	9
13	69	32	20	26	20	26	16	38	8
14	64	34	19	28	19	28	14	40	8
15	60	36	18	30	17	30	14	42	7
16	56	38	17	32	16	32	13	44	7
18	50	40	16	34	15	34	12	46	7
20	45	42	15	36	14	36	11	48	6
9 × 11	73	12 × 12	50	38	14	38	11	26 × 26	11
12	67	13	46	40	13	40	10	28	10
13	62	14	43	42	12	42	10	30	9
14	57	15	40	44	12	44	9	32	9
15	53	16	38	46	11	46	9	34	8
16	50	18	34	15 × 15	32	20 × 20	18	36	8
18	44	19	32	16	30	22	17	38	7
20	40	20	30	18	27	24	15	40	7
22	36	22	27	20	24	26	14	42	7
10 × 12	60	24	25	22	22	28	13	44	6
13	55	26	23	24	20	30	12	46	6
14	52	28	22	26	19	32	11	48	6
15	48	30	20	28	17	34	11	28 × 28	9
16	45	32	19	30	16	36	10	30	9
18	40	34	18	32	15	38	10	32	8
19	38	36	17	34	14	40	9	34	8
20	36	38	16	36	13	42	9	36	7
22	33	40	15	38	13	44	8	38	7
24	30	42	14	40	12	46	8	40	7
26	28	13 × 15	37	42	11	22 × 22	15	42	6
28	25	16	35	44	11	24	14	44	6
30	24	18	31	16 × 16	28	26	13	46	6
32	22	20	28	18	25	28	12	48	5
34	21	22	25	20	23	30	11	30 × 30	8
36	20	24	23	22	21	32	10	32	7
38	19	26	21	24	19	34	10	34	7
40	18	28	20	26	17	36	9	36	7
42	17	30	18	28	16	38	9	38	7
11 × 12	55	32	17	30	15	40	8	40	6
14	47	34	16	32	14	42	8	42	6
15	44	36	15	34	13	44	7	44	6
16	41	38	15	36	13	46	7	46	5
18	37	40	14	38	12	48	7	48	5
19	34	42	13	40	11	24 × 24	12	50	5
20	33	14 × 14	37	42	11	26	12		
22	30	16	32	44	10	28	11		

LINEAR EXPANSION OF SUBSTANCES BY HEAT.

To find the increase in the length of a bar of any material due to an increase of temperature, multiply the number of degrees of increase of temperature by the coefficient for 100 degrees and by the length of the bar, and divide by 100.

Name of Substance.	Coefficient for 100° Fahrenheit.	Coefficient for 180° Fahrenheit, or 100° Centigrade.
Baywood, (in the direction of the grain, dry,)	.00026 TO .00031	.00046 TO .00057
Brass, (cast,)	.00104	.00188
" (wire,)	.00107	.00193
Brick, (fire,)	.0003	.0005
Cement, (Roman,)	.0008	.0014
Copper,	.0009	.0017
Deal, (in the direction of the grain, dry,)	.00024	.00044
Glass, (English flint,)	.00045	.00081
" (French white lead,)	.00048	.00087
Gold,	.0008	.0015
Granite, (average,)	.00047	.00085
Iron, (cast,)	.0006	.0011
" (soft forged,)	.0007	.0012
" (wire,)	.0008	.0014
Lead,	.0016	.0029
Marble, (Carrara,)	.00036 TO .0006	.00065 TO .0011
Mercury,	.0033	.0060
Platinum,	.0005	.0009
Sandstone,	.0005 TO .0007	.0009 TO .0012
Silver,	.0011	.002
Slate, (Wales,)	.0006	.001
Water, (varies considerably with the temperature,)	.0086	.0155

MENSURATION.

LENGTH.

Circumference of circle = diameter × 3.1416.
Diameter of circle = circumference × 0.3183.
Side of square of equal periphery as circle = diameter × 0.7854.
Diameter of circle of equal periphery as square = side × 1.2732.
Side of an inscribed square = diameter of circle × 0.7071.
Length of arc = No. of degrees × diameter × 0.008727.
Circumference of circle whose diameter is 1 =

$$\pi = 3.14159265.$$

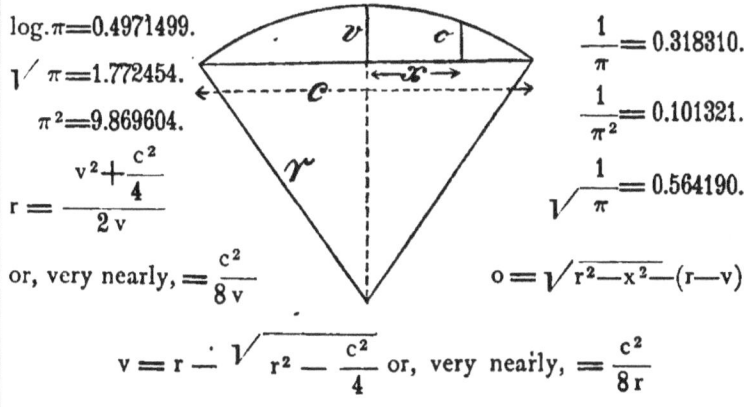

$\log.\pi = 0.4971499.$
$\sqrt{\pi} = 1.772454.$
$\pi^2 = 9.869604.$

$r = \dfrac{v^2 + \dfrac{c^2}{4}}{2v}$

or, very nearly, $= \dfrac{c^2}{8v}$

$\dfrac{1}{\pi} = 0.318310.$
$\dfrac{1}{\pi^2} = 0.101321.$
$\sqrt{\dfrac{1}{\pi}} = 0.564190.$

$o = \sqrt{r^2 - x^2} - (r - v)$

$v = r - \sqrt{r^2 - \dfrac{c^2}{4}}$ or, very nearly, $= \dfrac{c^2}{8r}$

AREA.

Triangle = base × half perpendicular hight.
Parallelogram = base × perpendicular hight.
Trapezoid = half the sum of the parallel sides × perpendicular hight.
Trapezium, found by dividing into two triangles.
Circle = diameter squared × 0.7854; or,
     = circumference squared × 0.07958.
Sector of circle = length of arc × half radius.

MENSURATION—Continued.

Segment of circle = area of sector less triangle; also, for flat segments very nearly $= \frac{4}{3} v \sqrt{0.388 \, v^2 + \frac{c^2}{4}}$

Side of square of equal area as circle = diameter × 0.8862; also, = circumference × 0.2821.

Diameter of circle of equal area as square = side × 1.1284.

Parabola = base × ⅔ hight.

Ellipse = long diameter × short diameter × 0.7854.

Regular polygon = sum of sides × half perpendicular distance from center to sides.

Surface of cylinder = circumference × hight × area of both ends.

Surface of sphere = diameter squared × 3.1416; also, = circumference × diameter.

Surface of a right pyramid or cone = periphery or circumference of base × half slant hight.

Surface of a frustrum of a regular right pyramid or cone = sum of peripheries or circumferences of the two ends × half slant hight + area of both ends.

The following formulæ are used to obtain the areas of irregular plane surfaces which are bounded by a base line, "cc," and two ordinates, "a" and "b," as per figure.

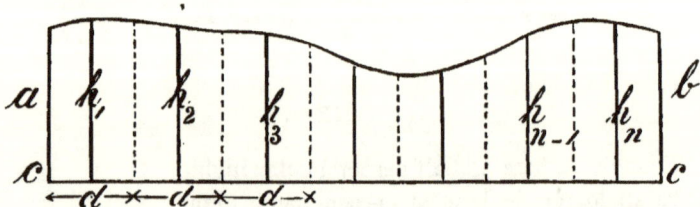

The formulæ are given in the order of their accuracy, beginning with the most accurate.

The surface is divided into any number (n) of parallel strips having the same widths, d, and whose middle ordinates are represented by $h_1 \, h_2 \, h_3 \, \ldots \ldots \ldots \ldots \, h_{n-1}$ and h_n

MENSURATION—Continued.

I. Area $= d \times \Sigma h + \dfrac{d}{72}(8a + h_2 - 9h_1) + \dfrac{d}{72}(8b + h_{n-1} - 9h_n)$

(Francke's rule.)

II. Area $= d \times \Sigma h + \dfrac{d}{12}(a - h_1) + \dfrac{d}{12}(b - h_n)$

(Poncelet's rule.)

III. Area $= d \times \Sigma h$.

These formulæ are more convenient for use than Simpson's rule, and I and II give generally and III sometimes more accurate results.

Σ stands for *sum of*.

SOLID CONTENTS.

Prism, right or oblique, $=$ area of base $\times$ perpendicular hight.
Cylinder, right or oblique, $=$ area of section at right angles to sides $\times$ length of side.
Sphere $=$ diameter cubed $\times$ 0.5236.
  also, $=$ surface $\times$ ⅙ diameter.
Pyramid or cone, right or oblique, regular or irregular, $=$ area of base $\times$ ⅓ perpendicular hight.

PRISMOIDAL FORMULA.

A prismoid is a solid bounded by six plane surfaces, only two of which are parallel.

To find the contents of a prismoid, add together the areas of the two parallel surfaces and four times the area of a section taken midway between and parallel to them, and multiply the sum by ⅙th of the perpendicular distance between the parallel surfaces.

WEIGHTS AND MEASURES.

AVOIRDUPOIS OR ORDINARY COMMERCIAL WEIGHT.

UNITED STATES AND BRITISH.

Ton.	Cwts.	Pounds.	Ounces.
1.	20.	2240.	35840.
0.050	1.	112.	1792.
	0.0089	1.	16.
		0.0625	1.

1 pound = 27.7 cubic inches of distilled water at its maximum density, (39° Fahrenheit.)

LONG MEASURE.

UNITED STATES AND BRITISH.

Miles.	Rods.	Yards.	Feet.	Inches.
1.	320.	1760.	5280.	63360.
0.003125	1.	5.5	16.5	198.
0.000568	0.1818	1.	3.	36.
0.0001894	0.0606	0.3333	1.	12.
0.0000158	0.005051	0.02778	0.08333	1.

The British measures are shorter than those of the U. S. by about 1 part in 17230 or 3.677 inches in a mile.

A fathom = 6 feet. A Gunter's surveying chain = 66 feet or 4 rods, 80 chains making a mile.

SQUARE OR LAND MEASURE.

UNITED STATES AND BRITISH.

Sq. Miles.	Acres.	Sq. Rods.	Sq. Yards.	Sq. Feet.	Sq. Inches.
1.	640.	102400.	3097600.	27878400.	
	1.	160.	4840.	43560.	6272640.
		1.	30.25	272.25	39204.
		0.0331	1.	9.0	1296.
			0.111	1.	144.
				0.00694	1.

WEIGHTS AND MEASURES—Continued.

CUBIC OR SOLID MEASURE.

UNITED STATES AND BRITISH.

1728 cubic inches = 1 cubic foot.

27 cubic feet = 1 cubic yard.

A cord of wood = $4' \times 4' \times 8'$ = 128 cubic feet.

A perch of masonry = $16.5' \times 1.5' \times 1'$ = 24.75 cubic feet, but is generally assumed at 25 cubic feet.

DRY MEASURE.

UNITED STATES ONLY.

Struck Bush	Pecks.	Quarts.	Pints.	Gallons.	Cubic Inch.
1	4	32.	64	8.	2150.
	1	8.	16	2.	537.6
		1.	2	0.25	67.2
		0.5	1	0.125	33.6
		4.	8	1.	268.8

A gallon of liquid measure = 231 cubic inches.

A heaped bushel = $1\frac{1}{4}$ struck bushels. The cone in a heaped bushel must be not less than 6 inches high.

A barrel of U. S. hydraulic cement = 300 to 310 lbs., usually, and of genuine Portland cement = 425 lbs.

To reduce U. S. dry measures to British imperial of the same name, divide by 1.032.

NAUTICAL MEASURE.

A nautical or sea mile is the length of a minute of longitude of the earth at the equator at the level of the sea. It is assumed = 6086.07 feet = 1.152664 statute or land miles by the United States Coast Survey.

3 nautical miles = 1 league.

COMPARATIVE TABLE OF
UNITED STATES AND FRENCH MEASURES.

MEASURES.	No.
One grain = gramme,	0.0648
One pound avoirdupois = kilogramme,	0.4536
One ton of 2240 lbs. = tonnes,	1.0160
One ton of 2000 lbs. = tonne,	0.9071
One inch = millimetres,	25.400
One foot = metre,	0.3048
One mile = kilometres,	1.6094
One square inch = square millimetres,	645.2
One square foot = square metre,	0.09291
One acre = are (100 square metres),	40.47
One square mile = square kilometres,	2.590
One cubic inch = cubic centimetres,	16.39
One cubic foot = cubic metre,	0.02832
One cubic yard = cubic metre,	0.7646
One quart dry measure = litres,	1.101
One quart liquid or wine measure = litre,	0.9465
One foot pound = kilogrammetre,	0.1383
One pound per foot = kilogrammes per metre,	1.488
One thousand pounds per square inch = kilogramme per square millimetre,	0.703
One pound per square foot = kilogrammes per square metre,	4.882
One pound per cubic foot = kilogrammes per cubic metre,	16.02
One degree Fahrenheit = degree centigrade,	0.5556

COMPARATIVE TABLE OF
FRENCH AND UNITED STATES MEASURES.

MEASURES.	No.
One gramme = grains,	15.433
One kilogramme = pounds avoirdupois,	2.2047
One tonne = tons of 2240 lbs.	0.9843
One tonne = tons of 2000 lbs.	1.1024
One millimetre = inch,	0.0394
One metre = feet,	3.2807
One kilometre = mile,	0.6213
One square millimetre = square inch,	0.00155
One square metre = square feet,	10.763
One are (100 square metres) = acres,	0.02471
One square kilometre = square mile,	0.3861
One cubic centimetre = cubic inch,	0.0610
One cubic metre or stere = cubic feet,	35.3105
One cubic metre = cubic yards,	1.3078
One litre (one cubic decimetre) = cubic inches,	61.017
One litre = quarts, dry measure,	0.908
One litre = quarts, liquid or wine measure,	1.0566
One kilogrammetre = foot pounds,	7.2331
One kilogramme per metre = pounds per foot,	0.6720
One kilogramme per square millimetre = pounds per square inch,	1422
One kilogramme per square metre = pounds per square foot,	0.2048
One kilogramme per cubic metre = pounds per cubic foot,	0.0624
One degree centigrade = degrees Fahrenheit,	1.8

STRENGTH OF MATERIALS.

ULTIMATE RESISTANCE TO TENSION
IN LBS. PER SQUARE INCH.

METALS.

	Average.
Brass, cast,	18000
" wire,	49000
Bronze or gun metal,	36000
Copper, cast,	19000
" sheet,	30000
" bolts,	36000
" wire,	60000
Iron, cast, 13400 to 29000,	16500
" wrought, round or square bars of 1 to 2 inch diameter, double refined,	50000 to 54000
" wrought, specimens ½ inch square, cut from large bars of double refined iron,	50000 to 53000
" wrought, double refined, in large bars of about 7 square inches section,	46000 to 47000
" wrought, plates, angles and other shapes,	48000 to 51000
" " plates over 36" wide,	46000 to 50000

Wrought iron, suitable for the tension members of bridges, should be double refined, and show a permanent elongation of 20 per cent. in 5", when broken in small specimens, and a reduction of area of 25 per cent. at point of fracture.

The modulus of elasticity of Union Iron Mills' double refined bar iron is 25000000 to 26000000, from tests made on finished eyebars.

Iron, wire,	70000 to 100000
" wire-ropes,	90000
Lead, sheet,	3300
Steel,	65000 to 120000
Tin, cast,	·4600
Zinc,	7000 to 8000

STRENGTH OF MATERIALS—Continued.

TIMBER, SEASONED, AND OTHER ORGANIC FIBER.

	Average.
Ash, English,	17000
" American,	11000 to 14000
Beech, "	15000 to 18000
Box,	20000
Cedar of Lebanon,	11400
" American, red,	10300
Fir or Spruce,	10000 to 13600
Hempen Ropes,	12000 to 16000
Hickory, American,	12800 to 18000
Mahogany,	8000 to 21800
Oak, American, white,	18000
" European,	10000 to 19800
Pine, American, white, red and pitch, Memel, Riga,	10000
" " long leaf yellow,	12600 to 19200
Poplar,	7000
Silk fiber,	52000
Walnut, black,	16000

STONE, NATURAL AND ARTIFICIAL.

Brick and Cement,	280 to 300
Glass,	9400
Slate,	9600 to 12800
Mortar, ordinary,	50

ULTIMATE RESISTANCE TO COMPRESSION.

METALS.

Brass, cast,	10300
Iron, "	82000 to 145000
" wrought,	36000 to 40000

STRENGTH OF MATERIALS—Continued.

TIMBER, SEASONED, COMPRESSED IN THE DIRECTION OF THE GRAIN.   *Average.*

Ash, American,	4400 to 5800
Beech, "	5800 to 6900
Box,	10300
Cedar of Lebanon,	5900
" American, red,	6000
Deal, red,	6500
Fir or Spruce,	5100 to 6800
Oak, American, white,	7200 to 9100
" British,	10000
" Dantzig,	7700
Pine, American, white,	5000 to 5600
" " long leaf yellow,	8000
Spruce or Fir,	5800 to 6900
Walnut, black,	7500

STONE, NATURAL OR ARTIFICIAL.

Brick, weak,	550 to 800
" strong,	1100
" fire,	1700
Brickwork, ordinary, in cement,	300 to 450
" best,	1000
Chalk,	330
Granite,	5500 to 11000
Limestone,	4000 to 11000
Sandstone, ordinary,	4000

ULTIMATE RESISTANCE TO SHEARING.
METALS.

Iron, cast,	27700
" wrought, along the fiber,	45000

TIMBER, ALONG THE GRAIN.

White Pine, Spruce, Hemlock,	500 to 800
Yellow Pine, long leaf,	630 to 960
Oak, European,	2300
Ash, American,	2000

Cast iron columns, and wrought iron, ultimate strength of....79
Channel bars, lithographed sections of...................5-8
    "   " explanation of table on properties of.......56-61
    "   " table on properties of....................64, 65
Circumferences of circles, and areas..................112-124
Columns, corrugated, lithographed sections of..............15
    " Keystone octagon, lithographed sections of........13
    "   "   " thicknesses and corresponding
        areas and weights............................77
    " Piper's patent rivetless, lithographed sections of....14
    "   "   "   " thicknesses and correspond-
        ing areas and weights of.....................78
    " explanation of tables on.....................73-76
    " cast and wrought iron, ultimate strength of........79
    " wrought iron, ultimate strength of................80
    " wooden, ultimate strength of....................81
Comparative table of United States and French, and French
    and United States measures....................166, 167
Corrugated and galvanized iron......................85, 86
Cover angles, lithographed sections of....................12

Decimal parts of a foot for each $\frac{1}{64}$th of an inch............87
Decimals of an inch for each $\frac{1}{64}$th.....................171
Deck beams, lithographed sections of......................4
    "   " properties of............................63
Deflection of rolled eyebeams under load 33-55
    " formulæ for special cases 61
Dove tail, lithographed section of......................22

Elasticity, modulus of, assumed in tables..................60
    "   "   " for eyebars.......................168
Expansion, linear, of substances by heat..................160
Eyebeams ....See Beams.

Fence Iron, lithographed sections of......................22
Fire-proof floors...................................83, 84
Flat, beveled, lithographed section of....................22
Flat rolled iron, weights per lineal foot of...............88-93
    "   " areas of 94-99

	PAGE.
Flexure of beams of any cross-section, general formulæ on,	60, 61
Floorbeams of bridges	133
Floors and roofs, general notes on	82–84
Floors, lithographed illustrations of	23–25
Foot, decimal parts of, for each $\frac{1}{64}$th of an inch	100–103
French and United States measures, comparative table of	167
Galvanized iron	86
Gas pipe, sizes and weight of	132
Gauge, American, for sheet iron	111
" Birmingham, " "	110
Girders, riveted, table on	72
Glass, window, number of lights per box	159
Grooved irons, lithographed sections of	21
Half T's, lithographed sections of	15
Hand rails, " "	21
Ice slides, " "	22
Inertia, moments of, for usual sections	61
See also tables on properties of beams, channels, angles, etc.	
Keystone Bridge Co.'s corrugated iron	86
" " " standard proportions for upset rods	126, 127
" octagon columns, lithographed sections of	13
" " " thicknesses and corresponding areas and weights	77
Linear expansion of substances by heat	160
Loads per square foot, for floors, roofs, etc	84
Logarithms of numbers	153–155
Materials, strength of	168–170
Measures, and weights, United States and British	164, 165
" " " comparative table of United States and French, and French and United States	166, 167
Mensuration	161–163
Modulus of elasticity, assumed in tables	60

	PAGE.
Modulus of elasticity for eyebars	168
Moments, maximum bending, to be allowed on pins	136

Natural sines, tangents and secants..................144–152
Notes, general, on floors and roofs.....................82–84
Nuts, sizes and weights of hot pressed square..............130
   "   "   "   "   "   hexagon............131

Obtuse angle, lithographed section of.....................12
Octagon columns, lithographed sections of................13
   "   "   thicknesses and corresponding areas and weights..77

Patent post iron, lithographed section of..................15
Pillars, timber, ultimate strength of......................81
Pins, bearing value of, for one inch thickness of plate.......137
   " maximum bending moments to be allowed on.........136
Pipe, wrought iron, for gas, steam or water................132
Piper's patent rivetless columns, lithographed sections of.....14
   "   "   "   " thicknesses and corresponding areas and weights of.........................78
Plastered ceiling, weight of.............................84
Plastering, limit of deflection to allow for................31
Post irons, patent, lithographed sections of...............15
Posts....................................See Columns.
Pratt trusses, maximum stresses in......................141
   " truss, diagram of................................26
Properties of U. I. M.'s eye and deck beams............62, 63
   "   "   " channels......................64, 65
   "   "   " angle irons....................66, 67
   "   "   " tee irons.........................69
   "   "   " star irons........................69
   " explanation of tables on.....................56–61

Riveted girders, explanation of table on................70, 71
   "   " table on...............................72
Rivetless columns, lithographed sections of................14
   "   " thicknesses and corresponding areas and weights of......................................78

	PAGE.

Rivets and round-headed bolts, weight of..................125
Roof iron, lithographed section of......................20
Roofs, loads and weights per square foot for.............84
Round bars, and square, of wrought iron, weights and areas, and circumferences of round bars................104–109
Rule for finding the area of a bar of wrought iron, given the weight, and vice versa............................84

Sash Irons, lithographed sections of.....................22
Screws, wood...129
Separators, between beams, lithographed illustrations of......24
    "   "   " weight of......................82
Shearing and bearing value of rivets....................135
Sheet iron, by Birmingham gauge........................110
    "   " American   ".......................111
Sines, tangents and secants, natural................144–152
Spacing of beams in floors............................33–55
Spikes, wrought..129
Square root angle irons, lithographed sections of............11
Square and round bars of wrought iron, weights and areas, and circumferences of round bars................104–109
Standard screw threads, nuts and bolt heads, recommended by the Franklin Institute...........................128
Star irons, lithographed sections of......................12
  "   " properties of...................................69
Strength of cast and wrought iron columns..................79
  "   " wrought iron columns.........................80
  "   " timber pillars..............................81
  "   " materials............................168–170
Stresses, maximum, in Pratt trusses......................141
  "   " in Whipple trusses..............142, 143
  "   explanation of tables on...................139, 140
Struts.. See Columns.
Substances, weight of a cubic foot of.................156–158
  " linear expansion of, by heat..................160

Tacks..129
Tee, half, lithographed sections of.......................15
Tee irons,   "   " 16–20

	PAGE.
Tee irons, properties of	69
Threads, screw, Franklin Institute standard	128
" " Whitworth standard	129
Tie rods, for brick arches in floors	83
Timber beams, safe load for	138
" pillars, ultimate strength of	81
Tubes, wrought iron welded, for gas, steam or water	132
Upset screw ends for square and round bars	126, 127
Weights of a cubic foot of substances	156–158
" angle irons, corresponding to thicknesses varying by $\tfrac{1}{16}''$	68
" brick-work, walls of	83
" flat rolled iron	88–93
" nuts, hot pressed, square and hexagon	130, 131
" separators and bolts	82
" sheet iron	110, 111
" square and round bars of wrought iron	104–109
" tubes, of wrought iron, for gas, steam or water	132
" wrought iron, rule for finding, given the area	84
" " spikes, wood screws and tacks	129
Weights and measures, United States and British	164, 165
" " " comparative table of United States and French, and French and United States	166, 167
Whitworth standard screw thread	129
Window glass, number of lights per box	159
Wooden beams, safe load for	138
Wood screws	129
Wrought spikes	129
Z iron, lithographed section of	22

www.ingramcontent.com/pod-product-compliance
Lightning Source LLC
Chambersburg PA
CBHW031452160426
43195CB00010BB/955